Mass Contacts

Stefano Breccia

authorHOUSE®

AuthorHouse™ *UK Ltd.*
500 Avebury Boulevard
Central Milton Keynes, MK9 2BE
www.authorhouse.co.uk
Phone: 08001974150

© *2009 Stefano Breccia. All rights reserved.*

No part of this book may be reproduced, stored in a retrieval system, or transmitted by any means without the written permission of the author.

First published by AuthorHouse   1/29/2009

ISBN: 978-1-4389-0678-2 (sc)

Library of Congress Control Number: 2008911602

Printed in the United States of America
Bloomington, Indiana

This book is printed on acid-free paper.

Foreword

A book on the "Pescara's legend"
revives a mystery that has slept virtually undisturbed for years

Italy 1956:
Surrepticious Mass Contact

The backgrounds of a troubling scenario

In his recent book "Accadeva a Pescara" (= "Incident in Pescara") Tullio Bosco has written: "In just the last few decades, research into parapsychology, and especially into telepathy, has become the object of careful studies. After weeding out clever magicians and conjurers that are able to cheat naïve audiences with their tricks, there remained persons, unfortunately uncultured ones, who possessed paranormal abilities but were unable to demonstrate them in convincing ways. Surrounded by sceptics, they were forced to give up any exhibits of their talents. To day, ESP subjects are eagerly sought in order to conduct systematic analysis of their skills. Through this process unexpected discoveries have been made. During the course of space missions around the Moon, research into telepathy has taken place. This new science is cultivated in the military under the strictest discretion. But back in 1946 ESP subjects were not only mistrusted, but were taken for delusionary, if not worse. Therefore, after their first experiences, these subjects often preferred to give up, resuming keeping their peculiar abilities secret.

That's what happened to young graduate Francesco C., son of a well-known manager from Pescara. He suddenly started hearing strange appeals, something like messages, emanating from a mysterious world. He talked about them with some friends, but nobody took him seriously. Francesco, nicknamed Ciccillo, was advised not to believe in dreams, but he was certain that they weren't dreams at all. So, one day, he made up his mind to follow up on those appeals, which were becoming more and more impelling. That night he got up, and, under the moonlight and winking stars, set off along the shore, toward Montesilvano (a small village North of Pescara). When he came to the edge of a long stretch of pines along the coast , he halted to glance at the northern sky. Before him stood the stately Little Dipper together with the wonderful North Star, reference point to so many ancient sailors. All of a sudden, just near the North Star, an elongated object appeared, with tapered extremities, flashing strange blue lights. The object was hovering, in the very middle of the Dipper. It wasn't easy to ascertain its distance, but it looked enormous. "A Flying Cigar!" the excited Ciccillo whispered, not being able to look away from the object. At that moment, a very smaller object, much brighter, came out, approaching at high speed. It was like an upside-down dish, rotating at a dizzying speed around itself, getting bigger and bigger. It stopped just in front of him, hovering a couple of meters above the ground, and then slowly descending onto the shore. Ciccillo was terrified for being so close to it, and at first he thought to run away, but that strange telepathic voice sounded again in his mind, telling him to stay put. In any case he could not have fled, for his legs were paralyzed. The flying saucer, about 15 meters in diameter, had a large dome at its center, surrounded with luminous portholes, and was sitting just in front of him, perhaps just for him! He was enchanted, unable to understand where those lights were coming from, those multi-coloured lights that were engulfing the craft, as if its motive power resided in them. All of a sudden a door opened, among the port-holes, and a little man got out. He walked up to the periphery of the disk, and raised his arms, in a friendly manner. The telepathic voice was now very strong, a message not framed in words, but of Ideas, of thoughts that Ciccillo was able to understand. "You will come here every time we call you. You will be the intermediary between us and men. Don't worry, we want to be your friends." Then the man returned inside the dome, the saucer started rotating again, lifted up, and disappeared among the stars in a matter of seconds. Newspapers reported the event estensively, due to its halo of mistery. Moreover, that same night, many flying saucers were reported by ships along the Adriatic coast, and some strange accidents occurred to boats in that area, which gave credence to the idea that the Adriatic sea had become a sort of Bermuda Triangle. The press expounded at length over these facts, and one of the most important magazines published a long article, with a picture of Ciccillo in it, as the hero of the unusual close encounter.

However, despite such coverage, Ciccillo felt that people were looking at him in a strange way, as if he didn't belong to this world. One day, possibly at the request of his relatives, a physician came to his home, supposedly just for a bit of chatting, but actually questioning him. It was plain that Ciccillo was no longer taken for a normal person. He started feeling closely watched, in particular when he was trying to go out at night. What could an ordinary physician understand of para-psychology? Although he was feeling more strongly than ever the lure of that mysterious world, Ciccillo was forced to repress his instincts, becoming more reserved, ceasing his usual friendliness, and no longer confiding even in with his relatives. Lack of understanding by the world, bitterness, sadness, and above all having been forced to give up a dream that had brightened his best years weighed heavily on the young man. If, on the contrary, everything had operated in his favour, would Ciccillo actually have been able to act as an intermediary between mankind and aliens? Nobody knows. Certainly, up till now the mistery about such beings is still impenetrable."

Bosco's booklet, apart from whatever consideration about what the author reports (it's evident that he was mistaken in speaking of 1946 – the correct date must have been 1964) re-created the so-called "Pescara legend", which is still being discussed in our own times.

In his recent work "Alien Base", top-level UFOlogist Timothy Good, quoting from material I had sent to the prestigious magazine "Flying Saucer Review" (translated by Gordon Creighton) recalls that in April, 1961, Bruno Ghibaudi, a well known and acclaimed newsman, who wrote about technical and space matters, and was in charge of a TV program about model aircraft, stated that he had shot many pictures of unusual flying objects along the shore near Montesilvano (Pescara). As Good rightly emphasizes, one of these pictures represents such an unusual object, that the idea of a forgery is simply preposterous. Other pictures also showed flying saucers, at times in formation, swooping low over the shore of Pescara.

Later on, Bruno Ghibaudi was to publish other exciting UFO pictures, for instance those shot by the painter Gaspare De Lama of a flying saucer over Milano, which made the front page of the mass-circulation "Domenica del Corriere" magazine.

Ghibaudi had previously been assigned by the RAI (Italian State television) to make a survey about the people who claimed to have seen flying saucers in Italy. At first, he approached the job without too much an enthusiasm, interviewing people all around the country, but soon he was faced with an unexpected reality. Witnesses were numerous and reliable, both those who reported having seen and even photographed these mysterious objects, and those people who were said to have collected specimens at UFO landing sites, and even those who were alleging to have been present at landings, and to have observed the activities of UFO occupants. There were even people pretending to have met the UFOnauts. Along with those, there were reports about threats and menacing statements made to witnesses, so that people had better to stay silent. Witnesses had often been forced to conceal their experiences, fearing for their jobs, their reputations, their relatives and even for their social life which was sometimes indiscreetly intimidated by officialdom. Ghibaudi was disappointed. Then the scheduled TV broadcast was cancelled without explanation by the RAI management. Ghibaudi then published his findings in magazines (for instance, the "Settimana Incom") and newspapers ("Il Tempo"). He deserves to be called the second public figure, after the pioneering efforts by consul Alberto Perego, to publicize UFO activity in after-war Italy, and to have, therefore, become a reference source on the subject.

But there's more to it than that. Some months after having published his pictures (they actually appear to have been shot by the late Giancarlo De Carlo), in January 1963 (in the magazine "Le Ore") Ghibaudi claimed to have been invited to a meeting with "space people", that is a meeting with aliens who were worming their way into our society, working among us, in touch with Earth people serving in logistics and general support for these visitors, beings who weren't so different in appearance from ourselves. Ghibaudi has been rather vague on details, limiting himself to relating only what they had told him. To start with, that human (Homo sapiens) form is widespread and prevalent (although at times with slight variations) throughout the

universe, and that we are the result of an interbreeding among some ancient cosmic colonizers, who became corrupted following planetary disasters which had altered our surroundings.

Ghibaudi said that his interlocutors had a benign character, and were willing to help us, in spite of our violent and immature nature, that they are discreetly visiting our Earth, from time to time, not being perfect, nor infallible, but far beyond us both in technology and morality. In these meetings, they spoke about our misuse of nuclear power, and the risk of a generalized atomic war, stating that in any case, apart from some specific intervention on their own part, they were not allowed to interfere in the evolution of backward beings like us. The truth is, any species must cut its own path, facing the consequences of its failures and victories. They said that we are not ready yet for a mass contact, which would serve only to knock us out of balance, without teaching us what we can only discover by and for ourselves.

After such statements, Ghibaudi left the stage. He had been a pioneer in spreading UFO information, and this last interview had been a kind of mystic testament. While the Italian movie industry was producing "Il disco volante" (= "The Flying Saucer") by Tinto Brass, with Eleonora Rossi Drago, Silvana Mangano, Monica Vitti starring, and with Alberto Sordi playing four different characters at the same time, the mainstream press isolated Ghibaudi following his last interview, so that he paid too high a price for trying to report the truth, and he quietly left journalism. As in Tinto Brass's film, a contradictory, sleepy and conformist society such as Italy's one of those times could not endure any bombshell revelations, not even one coming from beyond our planet. The times were not ripe, and it was thought better to hide one's head under the sand, as ostriches do. Later on Ghibaudi was to be elected President of the Italian Society for animal welfare, and, consistent with the principles exhibited by the aliens he had met, he was respectful of nature and its inhabitants. Disenchanted and reserved, Ghibaudi avoided any further involvement in the UFO scene, taking refuge in his privacy. (Author: in 2005 Ghibaudi was awarded an encomium by the CUN, the main Italian UFO research center, and I am currently pressing him to write a book about his experiences)

But with which environment had Ghibaudi gotten in touch?

On February 11th, 1969, I took the exam of General State Doctrine, at the "Cesare Alfieri" Political Science Center, at the University of Florence. I was the last student there, and, when the exam was over, I remained packaging my books, and, while doing so, I overheard a portion of a talk between the professor and his assistant, who was reading a newspaper: "Again this rubbish about flying saucers!". "Hear now, they aren't rubbish" – answered the professor – "On the contrary it is too serious a matter.". It was enough to grab my attention; I had passed the exam easily enough, and was probably not going to see the professor again. So I approached him, and said "Good, professor. If you have five minutes of spare time, I'd like to talk a bit with you, over a cup of coffee". "Right – he answered – But I'll pay for the coffee.". So we sat at a table in a coffee-shop, talking about UFOs for an hour or longer. My professor seemed well prepared on the subject. When I told him that I was the General Secretary of the CUN, there was a *coup de scène*: he asked me "Are you aware of the underground alien base near Pescara and of its logistics?" I stood aghast, and not because I didn't know anything about it, for Perego had written, covertly, about it in his books, and had told me many additional details. I was astonished that my professor was telling me that he had been informed by persons whose societal level was extremely high, important names from industrial, governmental, and academic settings. Although he didn't mention the name "Amicizia", it was obvious that he too belonged to this occult fellowship, which had been created in order to help aliens in their enterprises on our planet.  Two years before, reporters Franco Bandini, Giancarlo Masini and Bartolo Pieggi, to celebrate the first 20 years of UFO's, had published a series of articles in the "Domenica del Corriere" magazine. Beginning with the February 26th, 1967, no. 9; they had described Amicizia as a regular fraud, but basing their opinions upon totally false information (mistaking, for instance the acronyms FW3 for W56, and Y14 for CTR), writing in a lurid style, and omitting geographical details. Bandini himself was to replicate the very same text in his book "Il mistero dei Dischi Volanti" (the first book where appeared the name CUN). I recalled these articles to my professor, who answered "Journalists often write shits, and, when things go the wrong way, and an idea becomes too well known, their only escape consists of creating a debunking smoke-screen that may quiet any excessive public interest. It works better if they can enlist the help of other journalists, especially those overflowing with self-esteem. The desired result is then assured. You see, not everyone cares to stay involved when the environment is too hypercritical for the general public and even for the authorities …".

Many would have been upset by such words, but not I, who had already heard similar tales from Alberto Perego, in private conversations. When in 1963 he published his third book, "L'Aviazione di altri pianeti opera tra noi" (= "Airships From Other Planets Are Here"), and just before being sent to Brazil as the Italian Consul in Belo Horizonte, the diplomat had announced another forthcoming title, "Dirò tutto" (= "I'll Tell You Everything"). However it was never published. His previous book "Sono Extraterrestri" (= "They Are From Another World"), which had some interesting pictures in it, including one showing the inside of a flying saucer that had landed in Italy, he presented some hazy information, and new intriguing pictures, showing for instance a saucer resting on the ground near Milano, and a presumed alien pilot. The diplomat seemed willing to go on the offensive, joining activists who aimed to gain possible access to an underground base, that some believed to lie under the Gran Sasso, others under the Maiella (two of central Italy's largest mountains). "They're crazy" Perego complied; "They wanted to explode mines to open their way …"

But he soon changed his mind. "I cannot write down what I had hoped to." he confided to me. "Either we must transform UFOlogy into a revolutionary political movement in order to overthrow old establishments, or we acknowledge that we must speak only when necessary, whatever our ideas may be. I have been made to understand that it is not yet the time to tell everything …".

"And why?" I asked.

"Because otherwise everything would be messed up, and we would be ruined. The aliens themselves do not want such an outcome. Should we go that way, they would be of no help to us. They would abandon us in an environment that would shatter us, in which we would be treated as phonies, lunatics or day-dreamers.

Speaking to Mario Maioli (the first President of the CUN) during a UFO convention that Stefano Breccia had organized in Pescara, Perego said: "There is Pinotti, there is Breccia … two boys who are now brilliant university students with open minds. We must "cultivate" them, because they will be the future of UFOlogy. It will be up to them to take the baton and work incisively over the long term, to achieve the goals which elude us today…".

And now, after having passed my exam, I was listening to similar stories from my professor: I realized that he didn't know much more than he was telling me. In any event, my professor convinced me to apply to Professor Giacomo Sani, of the institute of Applied Sociology, to submit an experimental doctoral dissertation on the subject of UFO's. He said: "We must plead our "Cause". I wasn't too optimistic, but actually his judgement was correct, and my request was accepted.

"Coincidence", some will say.

The title of my dissertation was to have been "UFO's and their social implications". Unfortunately, just as I was finishing, Prof. Sani took a position in Berkeley, and so he left the University of Florence. Problems arose, because there was no eligible professor to be my tutor, so the dean of the faculty suggested I change the title and to widen the framework of my dissertation. It was re-titled "Dimensions of Escape Today", had a chapter devoted to UFO's, was refereed without any problem, and so I got my doctoral degree.

What I learned from this event was that if such topics were accepted in elitist and academic environments, and discussed without apparent problems, it was due to the importance and pervasiveness of that structure Perego had told me about, and how readily it was accepted at such high levels.

Do not talk of coincidences!

Actually the stage was, to say the least, disconcerting. Since 1956 a group of human looking aliens from far away stars had built a huge underground structure along the shore of the Adriatic sea. By compressing the spaces among atoms, they had been able to make a cavern, 300 meters high, and several kilometers long, for use as living accomodations. It wasn't their only structure. There were many others, even under the sea, but the position of this one in the very center of the Mediterranean Sea made it highly important. Its dwellers had been named W56, where the W stands for "Victory", and 56 refers to the year it all started. They had a counterpart that had been fancifully named CTR (from the Italian word for "Opposite").

Access to the W56 "base" was obtained via a kind of "teletransportation", dematerializing alien craft, together with their crews, and rematerializing them within the structure, that was hosting several groups of "Visitants", in its various installations. One can't but become disoriented when thinking of mastering technologies such as compressing matter and space-time, travelling quicker than light, without actually overcoming this speed, moving within a kind of "hyperspace", all concepts far in advance of our current scientific discussions on these topics in both Europe and the United States. But all that actually took place, and cannot be ignored.

Do not speak about coincidences!

We should note that Perego himself was receiving alien letters by teletransportation. They consisted of short notes which materialized in the open air, and, just as in what parapsychologists term "apports", they were warm. Perego did show us such items.

After having completed their own main base, the aliens got in touch with some persons in Abruzzo (in east-central Italy), who were to become their "fifth column" of logistical support. Managed by the late Bruno Sammaciccia, what would later be commonly called Amicizia had been born, a discreet branch composed of trustworthy people, meant to fulfil the requirements of the aliens, namely, supplying specific items, raw materials, and consumables, as well as supporting them in every possible way. The late coordinator was also in charge of being a sort of public relations man, as well as a catalyst for interactions with specific environments (the church, the military, manufacturers, the press, scientists, academicians, politicians) that could be of some use to the cause. Just a few of them got involved. In more than a few instances, the rejection of would-be candidates was to give birth to rumors that it was all a fraud, and nothing else. But this had been foreseen, and was looked after.

Well in advance of the later and highly controversial tales concerning Ummites, the better known Bahavians, and the misunderstood Elta V and UTI, little by little these aliens, the ones physically similar to us, now dwelling on our planet had been worming their way into our society, infiltrating themselves into undercover roles in Italy, both in the industry and in areas of similar relevance. Their organization is said to have been active also in France, Switzerland, Germany, Austria, Argentina, and the former Soviet Union.

Had some alien had the need to "go home" after a while, a deadly accident would be staged (for instance, an aircraft crashing into the ground), and in this way he would have been able to get offstage. Others, assigned to settle for a long time within our environment, had even gotten married with Earth people, creating new children and families. This was possible because they are totally compatible, from a biological point of view, with ourselves.

Their social organization was hierarchical, but not as strongly as among their counterparts, the CTR. It would be silly and superficial to state that the two factions were at war against each other. "Star wars" *ante litteram*? Not really so, after all. Skirmishes, perhaps. This is not to be wandered at. It seems that there are many offworld cultures, who aren't always in agreement among themselves, but that doesn't necessarily mean fighting. Our Earth has for eons been a center of interest and is visited by many extraterrestrial races, at times using different approaches.

It's an easy and quiet way of life to have yourselves worshipped as gods. It has been done many times in the past. And it looks that even today this practice is continuing, and can be seen, for instance with BVM (Blessed Virgin Mary) phenomena, and things like that. Even an unimaginative person like Adamski ends by attributing to his "Space Brothers" his own unsubstantiated and only mindly spiritual attitudes. But the W56's, although ecologists, pacifists and pantheists, were not partial to inappropriately placed mysticism. Of course, not all aliens are like them.

Among the peoples visiting our Earth there seem to be aseptic nature lovers who would look after the evolution of living beings upon our planet, while restricting themselves only to monitoring. Then there are "merchants", looking only for easy ways to get raw materials, ethnologists studying the evolution of different cultures on this planet, missionaries caring for our future from an ethical point of view, militarily-oriented people overseeing national structures, aiming to maintain inter-stellar routes, and to protect us from assaults from outside, and many more. In a word, a multi-faceted "fauna", a very heterogeneous one, cemented by a common will not to interfere too directly with our affairs. But that's not all.

It's easy to acknowledge that our offworld visitors wish to protect their own safety. Therefore they very often make use of "biological robots", which are human clones, built and used as workers, expendable to be lost, if necessary, to spare their own creators from harm. Even more, such androids should have been the automatic pilots, with a physical build representing at the very least, or, on the contrary, accurate copies of Earth people, imitated for various purposes. It is important to recognize in this viewpoint what one Philip Corso has stated, about Greys being subservient to "Nordic" aliens.

Do not speak of coincidences!

In order to monitor our environment, the aliens were making use of miniaturized probes, spherical or discoid in shape, true remote-controlled sensors that were able to record images, sounds, even the thoughts of the people being observed. Adamski has also referred to registering disks, or registering globes, that were probably the "Foo fighters" seen in 1944-45, and the bogies of the "Battle of Los Angeles", in 1942. Today we tend to believe that they are similar to the Balls of light so often reported in connection with agro-glyphs, but also in places like Hessdalen, Sassalbo and Caronia (A Sicilian village, in which unexplained phenomena have been reported for years).

Do not mention coincidences!

Connections with the men from Amicizia were formal ones, even brotherly. Among them some would have been honoured with the thrill of piloting an alien scout. But over the longer term, humans would be shown not to be up to coping with the challenges, disappointing the expectations on the part of their alien friends. But probably the latter had not been granting them unlimited freedoms, to prevent possible unpleasant surprises. At the end this resulted in a critical situation, worsened in 1978 by the reduction of the W56's area of influence in favour of the CTR's. The Adriatic flap of that year was the outcome; after that the W56 dismantled their logistic structures, and abandoned our planet in the eighties.

This is not the plot of a second-rate science fiction film. It's the tale of an experience lived day by day by persons that I myself have known and esteemed, the vicissitudes of which only a few were aware, proceeding with the strictest discretion, according to the requirements of the Visitors. Today we find so many disclosers in the States who tell something rather similar stories which recall the Amicizia of 1956.

Do not speak about coincidences!

Perego used to say to me that "… should anything surface, beyond what has already leaked out, either the lot of us will all be assumed to be crazy, or chaos will break out. Of course there will be no "Dirò tutto!" (= "I'll Tell You Everything") ... I realize that actions taken must be long term ones. Even governments stay silent, waiting for better times, and to protect themselves. In Italy it will be up to brilliant youngsters, like you, like Breccia, and a few more, to study, to graduate, to assume social roles and gain a prestige sufficient to prepare the public opinion in a painless way for the new alien reality. You have to master what aliens have taught us since 1956, and spread this news not as revelations, but by making use of logic, reasoning, science, and the right arguments against your opponents. In this way a growing awareness will spread from nothing, so that the Great Revelation can take place, when "They" think it is convenient. You will have to help to create such conditions, and the new-born CUN managed by my friend Mario Maioli is the right key to it."

"I think you are the right man to take Perego's place in Italy" a prophetic George Amamski wrote me in 1965.

In any event, for more than 40 years I have worked to inform the public in order to mold public opinion, both in Italy and abroad, and, together with other team-mates, to create awareness of the rules of the game and of the stakes involved.

Whether it is true or not, Amicizia has carried forward this 40-years-long scenario, much of what we are revealing today.

Do not speak about coincidences!

For instance, in 1976, I was paid an unexpected visit by a fine young woman, whom I was never to see again. She told me she had come for the purpose of transmitting a direct message from the Visitors, who were using her as a medium. The woman cautioned me to reverse the personal decision I had taken to leave my role within the CUN because of problems with my family and my job. On the contrary, she said, I could not but take on that responsibility, and that I hadn't even negun to support the cause of UFO's. Moreover, she made some suggestions about my third book (nobody was aware that I was writing it!) and exhibited telepathic powers, discussing details of my life that only my wife and myself knew about. She foresaw for 1978 a u-turn in my future as a UFOlogist, enumerating a set of important events that have all, actually, taken place, from my preferential contact with the late Joseph Allen Hynek, to a collective sighting of a UFO, to the future public distribution by way of newsstands of the CUN's bulletin "UFO Notiziario" (at that time available only to subscribers) , and many other detail of events that were to take place. She went on saying that in spite of various difficulties I was to encounter, I had any way to go on quietly, according to a particular program that she expounded to me. Not surprisingly, it was actually consistent with what Perego had told me in advance about Amicizia, and I am still adhering to such suggestions.

Do not speak about coincidences!

I am not here to act as anyone's advocate, but only to testify to events indubitably coherent and complementary with one another, certainly far from happening randomly.

Apart from Perego and Ghibaudi (who did not mention it), and the picturesque disquisitions by Bandini (largely groundless), I have been the only one to speak explicity about Amicizia, mentioning some names (at first in "UFO: Scacchiere Italia", then in its expanded new edition "Oggetti Volanti non Identificati") hoping to persuade the persons once involved in it to come out into the open, after so many years had elapsed. This hope has been fulfilled only in part, and only with me. The protagonists are scared by a possible reprisal from the CTR, now that the W56's defence no longer exists (Author: not quite true: in this English version of my book, a new testimony appears, that of Mr. Gaspare De Lama), and they fear not being understood and accepted by laymen and by UFOlogists themselves. Furthermore they look only for peace and oblivion at this point. The passing away of Bruno Sammaciccia has changed this situation, and because in his will there was the request that his story be made known, without causing problems to anyone, I acknowledged that is was my duty to contribute to the truth, as much as possible.

I also understand that many will be puzzled by the association of a reality like Amicizia to my life experiences, when I have dealt with UFO problems in a serious scientific manner. But, when true, such things must be told in any case, before or after. It is true that I have always attacked every instance of bogus alien contact, from Bongiovanni to Rael, but at times I have been confronted with situations that could not be dismissed with a trivial scepticism. And after so many years of silence, I could even be charged with not having told everything, or to have told only what is convenient for me, while withholding conscious knowledge of much more. Sure, but that's it.

There are also imbecilic and dishonest people who state that I (and the CUN together) are in collusion with secret services (false), due to our contacts with institutional environments and Italian intelligence. In any case, everyone is free to believe what he likes. That's not my problem, for each person will be able to work out his answer, whether right or wrong for himself. For my own part, I have already found it, and my behaviour reflects it. It is not for me to make public those private details public, which could not be proven beyond my personal witnessing.

Let us remember, in any event, that if the case for UFO's were to be tried in a court of law, the sheer numbers of witnesses would override any scientific objection, and would be sufficient to win the verdict for UFO reality regardless of the judicial level hearing the case.

Roberto Pinotti

(General Secretary of the Italian CUN)

THE BEGINNINGS

Foreword

In this book I am going to write diffusely about aliens and, as the title suggests, of the instances of massive contacts between them and terrestrials.

I must state for the record that I use the term "aliens" simply for the convenience of it; I do not pretend to maintain that such entities are actually originating from remote stellar systems, nor am I totally convinced about that. But I state that questioning the true origin of these entities is a futile exercise: be they from Mars, from Wolf 424, whether they spring from the Little People, or other dimensions, or whatever, for us nothing changes. For decades UFO scholars have distanced themselves bashfully from the possibility that the phenomenon could be caused by sentient beings. Hopefully such an attitude is getting more and more feeble as time goes by.

There have been many examples of influences made on our lives by Martians (please forgive my using this term, just to avoid using the too aseptic "aliens". It's one of my bad habits), starting with the Biblical Book of Exodus (which should be read from the beginning to end, despite its heavy style, for the many funny discoveries it contains), up to our present days, with its whimsical statements about the Strategic Defence Initiative, with fanatic movements spreading everywhere, and with the unexpected taking of sides by the Vatican.

I am not going to speak specifically about UFO's, unless by accident, but rather about a phenomenon of which UFO's may at most represent a footnote. That is, I am going to relate events that may be unknown to many, concerning the few cases of mass contact, in which dozens, if not hundreds, of people are abruptly confronted by some external reality, nearly all at the same time. Thorough coverage of such a subject would require that an encyclopedia-length book be written about it. However I decided to limit myself to the exploration of the most significant cases, leaving aside the ones which do not show sufficient plausibility. In doing this, I have sharply reduced the amount of material to be discussed, though I am certain that omitted material deserved its own detailed study, at least from a psychological viewpoint. I have also decided to ignore almost all the cases of individual contacts, thus reducing another significant part of the phenomenon which, again, deserves its own study but which is beyond the scope of this book.

Quite often I will speak about the deficiencies of our sciences, and of Science's inability to consider important information, because of the dysfunctional structure of Italian (and foreign) universities. Because of the profitability involved in presenting unfounded statements as revealed truth, the academic world is facing an unfortunate future, if they cannot dislodge a centuries-old culture of secular encrustments and political intrigues, and if they do not seek out a different way of financing themselves.

For instance, every New Year's Eve, the solons on duty come out, thundering against astrology and those who earn a living from it (to erase all doubts, I too haven't the least trust in such subjects), but conveniently forgetting to comment on how much they themselves are earning from similar totally insupportable concepts. Unfortunately our "scientific" world is full of big (actually little) men like that. In Italian universities politics is becoming more important than research, so that we are confronted with a very gray scenario. At times the Martians actually try to give us information that could allow our sciences to take a giant step forward, but unfortunately the vested interests seem to be more interested in completely different priorities.

A sad scene: some people (hopefully very few of them), engaged at that time with Amicizia, changed completely their minds; I am speaking about the so-called scientific luminaries, who, according to me, could not even be permitted to write stupidities in partisan newspapers and, at the same time, be proud of the academic honours, earned at times in an obscure way. One banal example: at the end of the sixties, I lectured at the faculty of Physics in the university of Bologna. On that occasion, I had strongly undermined the idea that it is impossible to go below the minus 273 and so centigrade degrees temperature of Absolute Zero. I was pretending that such a statement was totally unsupported, although at that temperature we register the end of molecular agitation. This means that the derivatives of spatial coordinates with respect to time become zero. And so? Whoever pretends that this temperature limit is actually absolute forgets the Riemann metrics (or, perhaps, doesn't even know what they are). Well, that night none of the professors listening to me was able to give me the lie. Nevertheless, in schools they go on speaking about the Absolute Zero.

And so on and so forth ... Being an engineer fond of physics and mathematics, I am obviously interested in such topics. But, as you are going to read later on, I do not shrink from making excursions into fields that are orthogonal to these disciplines. Indeed I've often been forced to do so by what our Martians were imposing on us. Further more, I will probably make use of the alien scenario in order to underline concepts that, apart from aliens themselves, should give to scholars something to think about.

Finally, I suggest that the reader examines what follows, with a fiercely critical attitude, but without any prejudice originating from the sciences, philosophy, religion, or any other source. I do not pretend to be the medium to deliver the information that has come from our Martians (I would not be able to do that), nor to be their champion. I will merely present the most significant concepts, obviously leaving the reader free to express his/her own judgement about my statements.

It's my duty to thank those who have helped me with this work, above all Prof. Bruno Sammaciccia, who had the general idea, then Eng. "Hans", who has supplemented the information coming from Bruno; also Dr. Galina, a Siberian scholar in physics; Giancarlo, who has been my partner in so many technological adventures; Prof. Paolo Di Girolamo, an old friend with whom I share many points of view, and reject many others; Dr. Roberto Pinotti, another long-time friend, who has persuaded me to travel this road. Then many other friends, who have been of help with suggestions and critics; among them, Eng. Carlo Bolla, who has always been nearby during the successive drafts, and to whom I owe several valuable hints. Among the most unforeseen ones: this text (Italian version) was completed in January, 2006, that is 50 years after the story had begun for Bruno Sammaciccia! In 2005 the CUN has celebrated its 40th anniversary, in 2006 Amicizia celebrated its 50th. Then, the last ones, but only on a time scale: another good friend of mine, although we have met just once, Robert (Bob) Girard (Arcturus Books Inc.), who volunteered to revise my poor English; even before this book was printed, the two major Italian UFO magazines ("Notiziario UFO" and "Area 51") started publishing articles about it. In the November, 2007, issue of the latter, Mr. Fabio Siciliano (unknown to me) hinted that "Akrij" (you are going to find this name later on) might have had a meaning in Sanskrit, and I have been able to discover it, so I owe him the suggestion; then, of course, another long-time friend of mine, Paola Harris, who urged me to go on with this project, finding for me the adequate interlocutors in the States, and giving suggestions on how to proceed in an environment totally unknown to me.

Italy, and strange flying objects

When, in the early 50s, people and newspapers started discussing the strange aerial objects being seen everywhere, the acronym UFO had not yet been invented. They were speaking in terms of "Flying Dishes", then, later, "Flying Saucers", and only much later about UFO's.

When in Italy the first study groups were born, the subject of their study was called "Clipeologia", from the Latin "Clipeus", the round leather shield of Roman legionaries, with a bulge in its middle, about two feet in diameter. The first, obvious, reason for such an etymon lies in the fact a *clypeus* and a flying saucer look very similar to one another. But, much more importantly, it derives from the large number of Latin historians describing incidents involving *clipei*. For instance:

> *Sub occasu solis, orbis clipei similis ab occidente ad orientem visus est perferri.*

> (At sunset, a globe similar to a *clypeus* was seen travelling from west to east)
> (Julius Obsequens: *De Prodigiis*, XLV)

A well known philosopher stated that:

> *Alii vero ignes diu manent nec ante discedunt quam consumptam est omne quo pascebantur alimentum. Hoc loco sunt illi a Posidonio scripta miracula, columnaque clipeiqude flagrantes aliaeque insigni novitate flammae. Quae non adverterent animos, si ex consuetudine et lege decurrerent, ad haec stupent omnes quae repentinum ex alto ignem efferunt.*

> (Actually, fires go on indefinitely, and do not go out until they have consumed their source. Such miracles have been described by Posidonius, as having occurred in that place, together with columns and *clipei*, illuminated by their flames, an event never seen before.

Such things would not have terrified people, if they were an ordinary occurrence. On the contrary, people have been terrorized, because of the suddenness of their appearance, and the lights they were emitting from the sky)
(Seneca: *Quaestiones naturales*, 7, XX, 2)

Another quotation from another philosopher:

Scintillam visam e stella cadere et augeri terrae ad propinquitatem, at postquam lunae magnitudine fracta sit, inluisse eo nubilio diem, dein, cum in coelum se reciperet, lampadem factam umquam proditur Cn. Octavio C. Scribonio consulibus. Vidit it Silanus procunsul cum comitatu suo.

(A spark, emitting from a star, has been seen descending towards earth, then to grow as large as the moon, lighting the night with its glare; then it returned into the sky decreasing its light. This happened under the consulate of Cn. Octavio and C. Scribonio. The proconsul Silano had witnessed the event, together with his committee)
(Pliny the elder: *Naturalis Historia*, 2, XXXV, 1000)

In Roman times there was an obvious reason for *clipei* circulating in Italian skies; the Roman empire was, let's say from around 1000 B.C. to 1000 A.D., by far the greatest and most important civilization in the world, and therefore it was most subject to the interest of alien visitors.

In 312 A.D., Costantine and Massenzio were battling each other to gain control on the empire. The time was just after one of the cruelest persecution against the Christians. The night before the deciding battle, Costantine, looking at the starry sky, noticed a "sign". According to what he told his biographer Eusebius, the sign consisted of the Christ monogram, and Costantine had grasped the message: *In hoc signo vinces*

(= "In this sign, thou shalt conquer"). It seems that Costantine ordered his troops to put this monogram over their standards. What is certain is that, the next day, he defeated Massenzio once and for all, and that, after having won control of the empire, he ordered a drastic change in its policies towards Christians. His mother, St. Helena, indeed became an active hunter of relics in Palestine, finding many of them (whether authentic or not, but that doesn't matter). The night vision beheld by Costantine had changed the future of Christianity, and probably that of humanity.

Of course, we have three possibilities here. First, Costantine completely invented the whole story (But why? Before this apparition he had shown no benevolent attitude towards Christians). Second, it was a direct intervention by the Divinity (it would have been the first and only attempt of such a thing). And third, it was a direct intervention into the history of mankind, made by who had the power to do so.

In most recent times, a strange ray of light was described in the "Historia" by Ghilardacci:

> *Alli vinti et il dì seguente di Luglio alle cinque hora di notte in Bologna, fu un grandissimo terremoto, che parea che il mondo tutto volesse ruinare ... Nell'aria apparve una trave di fuoco ardente, che con grandissimo spavento ne andava al ciel volando.*

(The twentieth, and twenty-first, of July, at five o' clock at night (The reckoning of hours was different from today: 5 at night would have been around 11 p.m. according to our use - Author) in Bologna, a very strong earthquake took place; it appeared that everything was to be ruined ... In the sky an fiery beam appeared, flying all over and terrifying people)

This episode should have occurred between July the 20th and the 21st, 1399.

More recently (1558), the goldsmith and sculptor Benvenuto Cellini wrote his autobiography "La vita" (= "My life"). Among other things, in this book he tells of a night, when fleeing on horseback from Florence, together with a friend of his, he was astonished to see an object flying high in the sky, and shouted:

Oh Dio del cielo, che gran cosa è quella che si vede sopra Firenze? Questo si era com'un gran trave di fuoco, il quale scintillava e rendeva grandissimo splendore ...

(My God, what's that big thing over Florence? It was like a large lighted beam, emitting a blinding light)
(Benvenuto Cellini: *La Vita*, Cap. 89)

Of course painters too witnessed these immanent presences (long before cameras were invented!); in the photo section I present several masterpieces by Italian artists, all showing strange and incongruous devices: The "Nativity" by Ghirlandaio, "Bacchus and Ariadne" by Tiziano, and "The Annunciation" by Carlo Crivelli. Concerning this last picture, I owe my thanks to George Philip Britney who, on his website www.alienufosecret.com presents a scanning from this painting better than my own efforts had yielded.

It looks, in any case, as though this country is still under observation from entities from beyond Earth. Close to our own times, it appears that during the fascist period a "mother ship" was observed, together with some flying saucers around it, somewhere in the North-Eastern Italy. I present here two letters, written on the letter-head of the Italian Senate:

Non ometterò alcun particolare, e questo potrebbe
sembrare una stranezza di un aeronauta,
una volta conosciuti quelli che vo-
lano più veloci dei venti. So per certo
che è stato veduto da altri piloti
d'aviazione, anche da quel Mari-
nelli che ha già fatto il rapporto che
è arrivato a mani di [illegible].
[illegible], già cinque anni da ciò, dopo che
quelle forme [illegible] passate [illegible] [illegible],
è stato visto con questi ancora fra loro
[illegible] [illegible] di lungo [illegible] metal-
lico, grigio o argento. Nel rapporto
del [illegible] [illegible] è [illegible]:

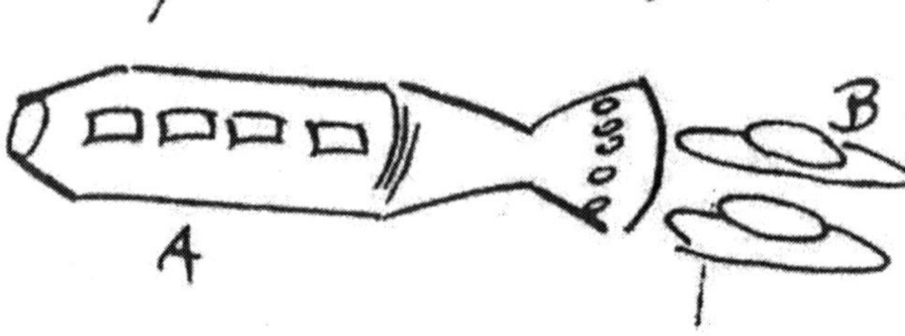

The drawings show a big "cigar", with a couple of Saturn-like objects around it. On the second paper there's also a sketch of an Italian fighter approaching. This letter, as well as following documents, were discovered quite recently, and until now it is unclear just what happened. It seem that "something" had landed. The Stefani Agency (practically the State press agency of those times) has sent at least three telegrams to every prefecture stating that the event was to be concealed under the strictest secrecy, by the order of the Duce himself.

Indicazioni di urgenza	UFFICIO TELEGRAFICO DI MILANO	Circuito sul quale si deve fare
PRIORIA SU TUTTE	TELEGRAMMA	===RISERVATISSIMO===

= = D'ORDINE PERSONALE DEL D U C E DISPONESI ASSOLUTO SILENZIO SU PRESUNTO ATTERRAGGIO SU SUOLO NAZIONALE AT OPERA AEROMOBILE SCONOSCIUTO STOP CONFERMASI VERSIONE PUBBLICANDA DIFFUSA DISPACCIO STEFANI ODIERNO STOP IDEM VERSIONE ANCHE AT PERSONALE AT GIORNALISTI STOP PREVISTE MAX PENE PER TRASGRESSORI FINO AT DEFERIMENTO TRIBUNALE SICUREZZA DELLO STATO STOP DARE IMMEDIATA CONFERMA RICEVIMENTO STOP = DIR GEN AFFARI SPECIALI = = FINE STOP = =

NO COPIA

RIS   RIS

2412

Mittente: Agenzia Stefani - Milano

UFFICIO TELEGRAFICO DI MILANO
TELEGRAMMA

=RISERVATISSIMO=

= = D'ORDINE SUPERIORE DISPONESI TRATTARE MODO SEGUENTE NOTIZIA DIFFUSA DISPACCIO STEFANI NR.66/3/1.C ODIERNO DUE PUNTI AEROMOBILE DI CUI SOPRA RICONOSCIUTO PER METEORA DICESI METEORA DA OSSERVATORIO ASTRONOMICO BRERA 500BP STOP DARE AT NOTIZIA MINIMA RILEVANZA GRAFICA STOP NON DICESI NON OCCORRE RETTIFICA STOP MINIMIZZARE STOP DARE IMMEDIATA CONFERMA RICEVIMENTO STOP = = DIR GEN AFFARI SPECIALI = = FINE STOP = =

NO
COPIA

Mittente: Agenzia Stefani - Milano.

Evidence shows that the government was upset. At first, newspapers were ordered to give no importance to this news, then to speak about a meteorite, then to actually melt the lead types used to print the story (in those times, there was no word-processing, and linotypes were still in use).

It was June the 13th, XI year of the Fascist Era, which is to say 1932....

It appears that, after this event, the government created a new Cabinet post, RS/33, devoted to studying innovative technologies, and headed by another inscrutable character, Guglielmo Marconi.

I am indebted to my old friends Robert Pinotti and Alfredo Lissoni, both from the CUN, for the research leading to the discovery of these documents.

Continuing into the fascist period, there must of course have been some reason behind this queer statement made by the Duce:

È più verosimile che gli Stati Uniti siano invasi,
prima che dai soldati dell'Asse, dagli abitanti, pare assai bellicosi,
del pianeta Marte, che scenderanno dagli spazi siderali
su inimmaginabili fortezze volanti.

(It is more likely that United States will be invaded
by the apparently very warlike inhabitants from the planet Mars,
who will descend from spaces in flying fortresses which we can't even
imagine, than by soldiers of the Axis's powers.)

(Benito Mussolini: Discorso alla Federazione Fascista dell'Urbe)
– February 25, 1941

Lord Cavendish, And His Group

Going a couple of centuries back, we find in Europe a sudden flowering of scientific discoveries, many of them quite out of character from the culture of those times. It makes one wonder about an early Amicizia, *ante litteram*.

I developed the suspicion that something strange had taken place while reading about Philippe Lebon (or Le Bon, according to other sources). This French engineer from the Department of Bridges and Highways, researching in a period when the science of Thermodynamic was firmly based on the phlogiston, discovered that the cold distillation of coal produced illuminating gas (a mixture of hydrocarbons, all with the formula C_6H_5OH). Thanks to him, the streets and avenues of the major European capitals ceased to be dark at night (the *Ville Lumière* among them). However, the discovery of illuminating gas was abruptly denying the theory of phlogiston. I wondered how Le Bon could have decided to undertake the experiments that were to lead him to his discovery (actually industrialized by Samuel Clegg, after Le Bon had passed away). Even more, how could the science of his times continue to supporting the previous theories about phlogiston, when it was now evident that such an hypothesis had been proven wrong.

In the XVIII century the history of sciences is full of similar examples. People had sudden, unforeseeable intuitions, often outside the scientific environment of those times, in an inexplicable way. What makes me think of an Amicizia *ante litteram* is that, although the characters involved were spread throughout central Europe, their focal point seems to have been a very eccentric individual, Lord Cavendish. But let us proceed in an orderly manner, beginning with some of the minor figures involved (minor indeed!).

Roger Boscovitch was born (by his own account) in Dubrovnik in 1711, became a Jesuit in Rome, authored, in 1736, a paper about sunspots, and was scientific advisor to the Pope. He restored St. Peter's dome in Roma, computed the meridians between Rome and Rimini, over a couple of degrees of amplitude, and explored the places where, a century and a half later, Schliemann was to dig up the remains of Troy. In 1760 he became a member of the English Royal Society. He died in Milano, in 1787. People like D'Alambert and Laplace declared themselves to be terrified by his innovating ideas, such as an unbelievable anticipation of the IBOZOO UU (as it is called in the context of the science of the Ummites – see later on), a concept that even today surpasses our mathematical understanding, and which is worthy of being studied, as this Jesuit did three centuries ago. During a long correspondence with Voltaire, Boscovitch discussed, for instance, about the transmission of malaria through mosquitoes, the possibility of other inhabited planets orbiting other suns, about something very similar to today's quantum mechanics three centuries before Planck, about Heisenberg's principle (again centuries in advance) and other similar trifles. Among other things, our hero had invented a new statistical algorithm that allowed him to confute Newton's hypothesis about the Earth being an ellipsoid! To this incredibly advanced genius an encyclopedia (I quote the first one I've found) devotes just four lines, dismissing him as the founder of the observatory in Brera! I am sure that today school-books do not even mention him.

At first, one would think that Pierre Ambroise Françoise Choderlos de Laclos (1741÷1803) should have been as remote from the sciences as possible. The fore-mentioned encyclopedia tells in a few lines that he has been a general under Napoleon, and a writer. This rude soldier, scarcely polished by the psychological introspections he explores in his novel (*Les liaisons dangereuses*), applied Newton's equations to the project of inserting an artificial satellite into a terrestrial orbit. The satellite was to be fired to the right height by a gun that had to make use of substances that our today's chemistry is not able to qualify as explosive, but which he claimed to have tested, two and a half centuries ago. Later on, Jules Verne discovered the project by Choderlos de Laclos in the annals of the *École Polytéchnique*, and made use of them in a well known novel. Unfortunately, he too had been a bit skeptical about the explosives quoted in that paper, and so he substituted gun-cotton for them. However Verne was not aware that the gun-cotton's explosive power was fully inadequate, so, sadly, his satellite would have fallen back to the grounds of the launch point in Florida!

Then there is the French Jesuit Louis-Betrand Castel (1688÷1757), whom Voltaire defined as "a *Don Quijote* of Mathematics". He is known as a precursor of synesthesia (and so I must consider myself an humble successor of his). Castel authored several works about mathematics, and the physics of gravity and light. In 1746 he was admitted into the Royal Academy in London, and, in succession, into the Academies in Bordeaux, Rouen, and Lyon. He wrote, as an evident antithesis to Newton, the *Traité de la pesanteur universelle,* in which he states that "... were it possible to get rid of the force of gravity, light itself should disappear, because the two things are intimately connected"! And this in 1724 ...

We then have François-Marie Arouet (1694÷1778), better known as Voltaire. In *Micromegas* he speaks about the two satellites of Mars (described in 1726 by Jonathan Swift in his *Gulliver's Travels*). He followed Bacon in epistemology, and Newton in Physics. In his *Letters on Newton* he explores in detail many astronomical problems and what was to later become the Calculus. Our present-day solons should reflect on his assertion that "Ignorance states or denies; science doubts".

Just a bit nearer to us, Sir George Cayley (1773÷1857) who, after Leonardo da Vinci, may rightly be considered the pioneer of flight. By the end of the 1700s he had already determined the fundamental parameters, the relationship between a wing's lift and its surface, how to compute the centre of pressure, and using a fin to increase the stability of an aircraft.

By 1796 he had designed a helicopter with two counter-rotating rotors and, *incredibile dictu*, had studied the possibilities of motorized flight, although the only motor available in his time, the one invented by Watt, was too heavy and too ineffective to be used on an aircraft.

Cayley was therefore forced to limit himself to gliders; he built many of them, and in 1849 he had a ten-year-old boy flying inside one of them. Later on his coachman was to experiment with another glider. It seems that, after landing, the man protested vehemently: "I have been hired to drive cabs, not gliders!"

Then, one more Jesuit (!), Bernard Le Bovier de Fontenelle (1657÷1757), the director of the French Académie des Sciences who, with the physics of his times totally devoted to Newton's equations, proposed again the apparently now obsolete theory of vortexes. He wrote about other inhabited planets, giving some examples (often very witty ones!), but also about the nature of stars and of comets. Moreover, our Jesuit was dissertating on Calculus, its applications to vortexes, and some other prettiness of this kind.

And finally, the most interesting personality among this group of far-sighted people, Sir Henry Cavendish. He was born in Niece on October the 20th, 1731, from a noble Norman-English line; his mother was Lady Ann Gray, daughter of the Duke of Kent, and his father was Lord Charles Cavendish, son of the second Duke of Devonshire. He endured a very poverty-stricken childhood, but, when he passed away (on February the 24th, 1810) he has left a really fabulous fortune. For instance, it was found that he had been one of the major stockholders of the Bank of England! Nobody has ever been able to understand where such a wealth had come from. Jean-Baptiste Biot, the French physicist, had to say: "He is the most wealthy among learned men, but also the most learned among wealthy people." He often used to lavish generously his money upon paupers.

Cavendish was actually a very whimsical person. He didn't like gatherings, he preferred to communicate through written notes instead of conversing; his voice was very high and strident, according to the few who heard it. He was apparently considering women to be a different race, and showed an almost pathological misogyny, to the point where his maids were to avoid to meeting him. If they failed to avoiding him, they were fired out at once. To make things easier for them, he had a staircase built at the back of his house, for their exclusive use.

Often he appeared to be unaware of this world. He once wrote to his housekeeper: "I would like for each of the gentlemen I've asked to supper to be served a mutton legs. I don't know how many legs a mutton has. Therefore it's up to you!"

Half an hour before dying he forewarned his housekeeper of his imminent passing away, and gave her his last will, among whose instructions was that his corpse was to be buried at once inside a simple grave, without any inscription.

He wore a long, torn, violet robe, a tricornered hat, and a wig with a pigtail, in the seventeenth century style. Only one portrait exists of this man, drawn behind his back, plus some caricatures.

Although having studied in Cambridge, he never earned a formal degree. Yet, when only 29 years old, he was admitted into the Royal Academy of Sciences, the first and only such case in the history of this prestigious institution, and in 1803 he was appointed as one of the 8 foreign associates of the *Institut de France.*

He gathered a vast library, probably the largest in his country, which he placed at the disposal of the academic world. He published almost nothing, and only more than a century after his death were some of his notes made known. It seems that even to-day nobody has confronted the task of examining in detail his many reports.

It may be recalled that although little is known of Cavendish's scientific research, he had computed the parallax of stars due to the mass of the sun long before Einstein, that in 1775 he had demonstrated to some of his friends the application of electricity, and the use of some instruments to make electrical measurements, formulating what were to be called, half a century later, Ohm's Law, and Coulomb's Law. Only in 1879 did J. C. Maxwell discover his studies on electrical condensers, and described them in his book *The Electrical Researches of the Hon. Henry Cavendish*, Cambridge University Press, 1879.

Thanks to the torsion scale he had invented, Cavendish has been able to evaluate the mass of the Earth, which he calculated as 5.448 g/cm^3 (compared to today's value of 5.515), and evaluated the universal gravitational constant, named after Newton. He primarily studied gases, in particular one, that he showed could be condensed into something similar to water, and that would easily burn; Lavoisier repeated the experiment, and named this gas Hydrogen (= "that brings water"). Cavendish showed that water is composed of two gases, and was able to synthesize water from hydrogen, oxygen, and an electric spark. Moreover he studied the noble gases, particularly neon and argon (strangely enough, the explosives devised by Choderlos de Laclos were compounds based upon these gases!). In 1784 he organized an ascent by aerostat, during which the two persons on board carried with them several bottles full of water. From time to time, they were to empty one of them, and to close it again; with this simple method, he collected samples of the air at different heights.

In short, during the XVIII century, central Europe has hosted many scholars, most of them ignored in the history of science, men who were making discoveries substantially incompatible with the culture of their times, if not far in advance of them. Fortunately for humanity, all of them had been in touch, in one way or another, with Sir Henry Cavendish, this focal point for men of talent who would otherwise have been left to themselves. Cavendish himself, we have seen it, was a man out of place in his times, and his intuitions incompatible with those of his peers. Indeed many among the scientists of his epoch had ignored his ideas.

It is certainly stimulating to think that Cavendish and his group may have been fed with suggestions *cum grano salis* by "someone" who had a wider vision of science, and who, perhaps, had chosen this method of sharply accelerating scientific development in Europe.

Later, in Fascist Italy, we had the Cabinet RS/33, and in Nazi Germany highly advanced researches into non-conventional weapons, most notably with Wernher Von Braun who was designing those engineering jewels that were the V2 rockets, and even inventing, from nothing, in a very short time, the technology of ceramic sintering which was necessary to manufacture their peculiar type of nozzle. That's another mystery in the history of technology: a country, already engaged in total war, creates from nothing a difficult technology, in a field of science about which, we must emphasize, both Von Braun and his colleagues were totally ignorant. Years ago, I was violently criticized for a paper in which I praised these rockets (of course, strictly from an engineering point of view!).

Once again, we have the feeling that, in one way or another, a technology transfer had taken place, on the part of "someone" who had the conceptual knowledge, the ability and the will to do so, on behalf of the backward human species.

More recently, William Shockley, John Bardeen and Walter Brattain were awarded the Nobel prize for inventing the germanium transistor while having been studying, at the Bell Labs, silicon diodes! And, at more or less the same time, optical fibres were developed, when the ancillary laser technology, needed to give them an application, was not yet available!

Up to now we have been following a path from the Roman Empire to our own times. We could have started before, and surely "hot" subjects for contemplation have been omitted, for instance the enigmatic character Leonardo da Vinci, looking for "strange" or curious study material. In very recent years, talking confidentially with friends that I can't afford to name, I found that some among the most quoted Italian intellectuals at the beginning of 1900 were engaged in something very similar. The story is evidently continuing, without a let-up.

Experiences in the States

In modern times, the best known forerunner by far among flying saucers contactees was George Adamski, a Polish refugee whose family emigrated to the United States when he was very young. He served bravely during the Mexican Civil war, and was eligible for burial in Arlington National cemetery. In the book he co-authored with Desmond Leslie, "Flying Saucers have landed", Adamski tells about his encounter with a "man from Venus" in the California desert. Some witnesses were present, who later signed affidavits confirming the facts.

To be sure, it was a very strange encounter. The Venusian seemed to be unable to understand English, and the two parties were forced to resort to using a whole range of *media*, from gestures to drawings, to some form of telepathy. An amusing twist to the story was the fact that his cosmic interlocutor had left Adamski a message on the ground… impressing esoteric-looking symbols sculptured on the soles of each of his shoes into the loose desert sand! In his "Other Tongues, Other Flesh", George Hunt Williamson, a key early saucer contactee in his own right, devoted the most of the book attempting to interpret these symbols. Even funnier, only a few weeks later, while meeting again with the same extra-terrestrials, Adamski was to find that his Venusian was now fluent in English! So, why such a parody?

Even more: In Adamski's tales, the Venusian was but one of the many extra-terrestrials that George was to meet, among whom not a few were … living in our society, were driving cars (It was typical of them to pick up the non-driving Adamski at his hotel, and to bring him back after a quick spin to the moon or to their mother ship), looking like normal businessmen.

According to Adamski, and to his biographers, such scenarios were not limited to the States; Louise Zinsstag (see Bibliography) reports an encounter with two aliens in Basel, who probably were able to speak German fluently. Strangely enough, in another book of hers, "George Adamski: Their Man on Earth", this incident doesn't appear.

Obviously, at a time when critics were scoffing in the press about "Little Green Men", the description of aliens so similar to ourselves was a bombshell. Nobody would have expected such an idea. That's the main reason why I hold that, on balance, we may trust Adamski. Of course he has depicted "his" aliens as totally adherent to his own philosophic theories (or perhaps it was their choise to present themselves as having such attitudes). Nevertheless Adamski never spoke about little green men, especially not with antennae! Something similar is taking place today. For the most part, people now wonder about bulb-headed Greys (which to me, even supposing that they do actually exist, are but some kind of biological robots), while the idea of human-looking aliens has been almost lost. But this will be the topic on which we shall concentrate - of "aliens" which in most cases would be impossible to detect as being anything but normal humans, often living completely at ease in our everyday world.

The technical and scientific setting that the aliens had presented to Adamski was just at the level the scantly-educated Pole was able to accept, often without understanding it. For example the aliens spoke of twelve planets within our solar system (a very old theosophical concept, here made preposterous by the comparison between the solar system and a cathode ray tube. Obviously Adamski's knowledge of geometry was very limited). Even more absurd the control panel of the scout ships: rows upon rows of push-buttons (the hands of a pianist, ideally with a span of half a meter and with ten fingers per hand would have been necessary to operate it), screens with some mysterious graphics animated on them, a lens embedded… on the floor, and so on. He was presented with something his uncultured background was hardly able to grasp (perhaps he had seen an oscilloscope somewhere, and maybe he had sometimes activated some switches at Palomar Gardens).

On the other hand there was the actual alien reality: light coming from nowhere, "space fireflies", and the revelation of woods and animals on the Moon. I believe that Adamski was introduced to a real alien environment (consider for instance the lighting), but the aliens, relying on their superior technology, showed him things totally fabricated for the occasion, without Adamski realizing the trick, at least at first.

Thanks to the deep scientific ignorance on the part of George Adamski, a tale has been peddled that no swindler would have dared to write in a book. In the last chapter of "Inside the Space Ships" Adamski describes walking in the vacuum of space, across the top of a mothership, without wearing a pressure-suit or any kind of protection! I am grateful to my friend Carlo Bolla for this reminder, because I had read the book many years ago, and had totally forgotten the episode. Today, thanks to information originating from within Amicizia, we are aware that such an exploit is really possible, and a minimal grounding in general physics can't but confirm it. Many pictures of the Space Shuttle should suggest us that many things do not conform to the physics we have been taught. Adamski was certainly not prophetic in this instance, he didn't realize the apparent absurdity of his tale, and simply wrote it down.

Some years ago, I had studied the photos of motherships which Adamski had inserted into his books from the standpoint of optical geometry. I realized that not only were they plausible, but that it was even possible to estimate the external dimensions of the object. It turned out to be a rather small ship, some tens of meters long; The study is appended to the end of this chapter. Those who pretend that the pictures show some distant galaxy would have been better off keeping their mouths shut!

Then, in Adamski's books, we find the "masters", who typically utter the most trivial banalities, and yet are listened to most carefully by all the people around, Adamski of course among them - a true absurdity.

Even more absurd is the tale of his trip to Saturn. Of course it's an easy job to dismantle it as an invention on Adamski's part. But, considering Adamski's behaviour before and after the supposed trip, I have the feeling that, once again, "someone" has been joking with him, an easy trick, after all, given their technical level, a trick intended to confirm him in his role of future spiritual guide of Earth, with Jesus Christ and other gurus applauding the mission.

And, whoever acted behind the stage, he got his desired result: although Adamski is the least plausible contactee, he has been the one with the widest visibility worldwide, with his visits to the Queen of Holland and to the Pope confirming him in his role. Whoever was behind these adventures, certainly showed a keen understanding of human psychology.

The other contactees of this time, in the States, are rather different from Adamski. Just one year after "Flying Saucers Have Landed", George Hunt Williamson published his "The Saucers Speak!". It is a text less engaged with theosophical concepts, and is a much more positive one than Adamski's.

The Italian edition was affected by a huge mistake by the translator: the sentence "To the apples we salt we return" has been rendered as "Mettiamo sale sulle mele e torniamo" (= "we salt our apples, then get back"). It would be unfair to saddle poor Mimi Robutti, the translator, with this mistake: how could she have understood the meaning of the sentence, that Williamson was to explain only in a later book? In any case, while I was still young, because of this mistake I conducted a lot of experiments involving apples and salt, which are better off being forgotten!

Although not the first book of its genre, "The Saucers Speak!" is, from my point of view, the first serious book about contacteeism, and moreover the first one to bring up the subject of mass contact. The technology was still rudimentary (we will discuss that in a greater depth later), but wireless was already in use. Early UFO contact were typically mono-directional. However, the reactions from bystanders can be recognized. For such reasons, I believe that this book is a UFOlogical classic – so much so, that when it was offered for sale by Robert Girard, a specialist in UFO literature, I bought the manuscript coy, paying a king's ramson for it, so that I now possess the original of the affidavit sworn by the witness to the contact.

There is something strange about this manuscript. It is type-written, with only one hand correction to the entire text; another (unpublished) manuscript of Williamson which I own is full of mistakes, corrections, and alterations, as a manuscript should be.

Another comment: in "The Saucers Speak!" only seldomly do the Williamsons play any significant role. It is Bailey, his wife, and the other persons who signed the affidavit. Therefore, it appears that, while Bailey and the others were engaged in the uoija board and radio methods of contact, George Williamson was already heading in different directions. Indeed, in a few years, his two subsequent books were to show depths of information that could not have been achieved through those sporadic contacts in Morse code. Actually, during one of the conventions held every year at the Giant Rock, Williamson played live a long spoken recording of a person allegedly from Maldek [1], supposedly referring to the destruction of the planet. It was a long strain on the audience's patience, and obviously incomprehensible, because the being was speaking in the Solex Mal tongue (an ancient galactic language described at length in Williamson's "Other Tongues, Other Flesh")! Evidence points to it being a radio recording superimposed over a conventional broadcast, in a way very similar to what was going to take place in Europe, just a few years later (actually, in the European cases, technology was usually better, so that the background broadcast was of no bother). Of course, the truthfulness of what the Maldek speaker was saying - in an unknown language, no less - cannot be verified, but the technical aspects of the broadcast are compatible with similar examples of this form of contact.

The experiences reported in "The Saucers Speak!" culminate in the unsuccessful attempt to arrange a physical encounter with their aliens. An encounter actually does seem to have taken place a couple of months later, with Adamski, Williamson and their companions being just witnesses, on November the 20th, 1952, the famous "Orthon" meeting.

Aside from its purely historical interest, I believe that the most interesting chapter of this book is the last one, titled "Saucers Still Speaking". Williamson tells how "... many groups are now experimenting in radio ... Some of these groups have had success." ... "We have had more personal contact recently with the saucers, but this is of such a nature, that we cannot put it into print, just now."

Some technical details are particularly interesting, because they were found to be quite similar to our European experiments. "I made some tests, and found the signal came from a certain direction, <u>straight to the radio</u>, NOT through the aerial" (underlining and capital letters are the Author's). And it ends with:

"Yes, the Saucers Are Still Speaking! Let's listen to what the have to say!

Our government, evidently, is not going to give us the information they now have. If they refuse us this knowledge, there are those of higher authority who will see that we are told!"

(For unknown reasons, this last statement is not present in the Italian version).

Among the persons in touch with Williamson, there was another radio ham, Robert Miller. He too was pretending to receive radio transmissions from aliens, and to have direct encounters with such beings. He reports a funny adventure that took place on October 24, 1954, somewhere around Detroit (see "Other Tongues, Other Flesh"). Miller was in his radio shack, waiting for some friends. All of a sudden, a message came through the set, ordering him to go immediately to a place nearby. He left at once, and at the place he found a flying saucer waiting for him. He got aboard, and, via wireless, he had a spoken exchange with his friends who, in the meantime, had arrived at his cabin! There are five signed witnesses also to this case, but what I consider relevant is the technology. Adamski had been given a mysterious message by way of the soles of his Venusian's shoes (!); Bailey and others, in the beginning, had, at first, communicated by Ouija Board, and only later by wireless in Morse Code Morse. But Miller, less than one year later, was now speaking by wireless directly to his friends. It's a bit difficult to believe that the alien wireless technology had evolved so rapidly, following a beginning from a technological level below our own! It is much easier to speculate that such limitations had been intentionally imposed by the supposed aliens, with some mysterious purpose in mind.

Later, in 1961, Williamson went mad: under the nickname of Brother Philip, he started propagating messages which, to oversimplify, we could call theosophical (it deserves a better clarification, but I do not believe this to be the right place to do that). Before that, in 1958, he had written "UFOs Confidential", together with McCoy, one of the first books (perhaps the very first one) to expose an occult governing of our planet. Therefore our anthropologist, the self-proclaimed rightful heir to the Serbian throne, had evolved from really reputable-looking contacts with aliens to Nazi-like concepts, along the lines of the "Protocols of the elders of Zion", only to wander into theosophy, or something like that. A wide change of attitudes, on the part of Michel d'Obrenovic (a.k.a. George Hunt Williamson).

Other contactees of that time were in general not so extreme. Many appear as unassuming, salt-of-the-earth types, often yanked from their mundane realities and commandeered to do the bidding of vastly advanced god-like figures from space. Occasionally funny situation develop (another clue that there must have been some truth to them). Bethurum, a Caterpillar mechanic (and so probably not very fluent in celestial mechanics), described the planet Clarion, a planet orbiting behind our Moon (!), therefore always invisible to us! Once again however, before starting to laugh, let's think for a moment about what this means. I do not believe that Bethurum was even aware of what "orbiting" means [2]; but not only did he maintain steadfastly such an absurd statement, but, in the mid-50's, he spoke of "Aura Rhanes" the beautiful woman captain of one of the "scows" coming from Clarion. In those days feminism was yet to fill the air with the smoke of burning brasseries, and, apart from a few significant exceptions (for instance Grace Murray Hopper, who invented the COBOL programming language, even today the most widely used language), it was difficult to find in the States (or everywhere else) a woman in charge of anything, even more so down at the blue-collar level. Nevertheless, our Caterpillar driver was adamant in championing this role for "his" Aura Rhanes. She liked to walk quietly along our streets, as she could not be distinguished from our women.

Bob Girard, while reviewing my text, at his point, sent me his opinion on Bethurum, which I am happy to include here, with his permission:

"He was my favorite for a long time, because he was so utterly PRIMITIVE as a human being, so incredibly dumb that he could not possibly have invented the story of his 11 encounters with the Admiral Scow and Aura Rhanes. I thought that for many, many years. But then I gradually understood that Aura was a TULPA, created BY Bethurum out of nothing, as a way of coping with the hideous life he spent under bulldozers and trucks by day, getting himself covered with dripping motor oil and brake fluid, and then trudging home at night to his equally hideous wife, Mary (read "Aboard a Flying Saucer" carefully - you will understand how she crushed every last atom of his almost non-existent remaining self-esteem, and absolutely hated his talking about flying saucers, threatening him with divorce, etc.). This had the effect of driving Bethurum even more quickly TOWARD flying saucers and their crews as a replacement life for a real world which could not have been more excruciating to any human being who ever lived. I once owned Bethurum's own scrapbook, which contained many newsclippings and other paper souvenirs of his saucer & literary career. To my astonishment, he had dedicated, in his own hand, on the inside front cover of the scrapbook, the whole collection TO Aura Rhanes and the rest of the Admiral Scow crew, in memory of those 11 visits. He truly believed in the reality of his better world, Clarion"

Some years later (and with Amicizia already running) Howard Menger tells funny details of his role of supporter to "his" aliens (as usual, people from Venus, from Saturn, and so on). I like to quote some extracts from his book:

"Such tasks were not without their moments of humour, and I think the visitors enjoyed them as much as I did. I remember one time when I was asked to purchase several complete outfits of female clothing ... I bought what I thought was the appropriate sizes and showed up at the point of contact. The women went into the next room, from which I soon heard a series of giggles and groans. Finally the door opened and the bras were flung out. They apologized, saying they just could not wear them, and they never had ... and you may be certain that I felt it wise not to ask! ... Once again they had difficulties – the high heels. They teetered and wobbled and suffered, but took it in good humour ... "Why can't your women wear sensible shoes!|" ... A man, with long blond hair, which hung to his shoulders, approached me and handed me a pair of scissors. He could not yet speak our language, as many of them couldn't at their first arrival ... This man simply pointed to his hair and sat down, and I presumed rightly that I had unwillingly become a barber! I took a handful of the finely textured hair and opened the scissors. I halted and looked at the man. His plaintive look made me feel sorry for him, for it was obvious he was proud of his beautiful hair ... He laughed and motioned to go through with it. I remember several occasions when I cut their hair. I don't know if they save the hair or not, however all evidence of the meetings was always carefully gathered up by the space people before they departed."

His aliens were apparently very realistic. Menger has given us among the best photos of the bell-shaped, Adamski-style scout ships, with aliens visible nearby, beautiful movies, plus some fakings – it happens very often, unfortunately). Just a footnote to the abobe: Adamski style bell-ships have almost never been seen, since the days of Howard Menger. Since then only a few have been reported, mostly in Abruzzo and Marche, in Italy, and in Munich in Germany.

Menger's aliens often sought his help in order to blend with Earthmen without being noticed. It strains credibility to think that beings with such advanced technology, could be in need of a local barber. Menger was earning his living as a sign painter, so he was probably not much more culturally polished than Bethurum. He went along willingly enough with the errand-boy routine. Yet, in his case, there were several witnesses to support his claims. In all likelihood, "someone" had led him to believe that his role was vital to the successful outcome of some highly important extraterrestrial mission. This someone had used Menger in a long series of contacts, allowing the "plot" to end there or to be carried on from there, and had rewarded Menger with continuous contacts and exalted his credibility by the way of meetings with aliens in front of different witnesses (of course, I must rely solely on his book for this viewpoint). Then, suddenly, everything ended.

Much later, we find Daniel W. Fry, Ph.D. as he styled himself. Generally speaking, Fry was an aeronautical engineer who underwent a series of experiences with an alien, by the name of Alan. His book "The White Sands Incident" also shows a number of photos that, although at times disputable, represent nonetheless a new type of alien ship.

Fry's point of view is more similar to my own, although, strangely enough, he never emphasized significant details. The only comment I wish to make about Fry is that, once having started to read his "Atoms, Galaxies and Understanding", I stopped quickly, thinking that it was a simple treatise on elementary physics. However, much later I started to read it again, but this time found (after the point where I had left it) a noteworthy treatise on contemporary physics, surprising for the ease with which he explores some really difficult concepts. Interestingly enough, "Atoms, etc" was published in 1960, while the classical "The White Sands Incident" was six years earlier.

Fry's contact seems to be limited to himself alone, and therefore we will leave off consideration of him (if I were to discuss all the contact cases I know, I'd have to write an encyclopedia!).

I could also write an encyclopedia if I wrote about all the mass contact cases, true or false, of which I am aware. The so-called New Age is full of such instances, and it is difficult to acquire information on each case. What is of interest to me is that between the mid-50's and mid-60's the United States was the scene of one significant instance of a mass contact (Williamson *et alii*). That is, a situation in which a number of persons, connected or not with each other, where in touch with the Brothers from Venus or elsewhere. Both in mass contacts and in individual ones (i.e. Adamski), Earthmen were confronted with human-looking beings who were able to walk our streets without being noticed as "alien".

In coming chapters, we will examine a couple of such instances. One is rather well known (the UMMO Affair, but I will present information of which many people are unaware), while the second one has been until now unknown by almost everybody, but yet is probably the most important case known in human history. I'd like to underline that it shows significant differences from the others: an extremely wide geographical theater (Europe, Siberia, Argentina, Australia), a very long duration (several decades), a strong sense of orchestration in its operative phases: a high philosophical concentration, no whimsical sermonizing, and above all, a strong attention to immanent details, creating many problems for the Earth people involved. It is one thing is to discuss (perhaps inside a flying saucer) high level problems. It is quite another thing, however, to arrange for a couple of big lorries, full of fruit, twice a month. Adamski and Fry had mentioned mercury, but it is much more challenging to acquire industrial quantities of this strange metal, together with strontium niobate of barium, platinum and radium (in times when the sale of radio-active materials was illegal both in Italy and in the most of the world). One may propagandize about the astonishing properties of some particular compound, but providing and testing them is a totally different thing. Trying to put technological information from alien sources into practice is a daunting task, requiring the spending of a great deal of time and of money in the process, while at the same time refusing interested solicitations on the part of the major powers, and doing one's best to shut the door in their faces. To risk being burned because an experiment got more powerful results than expected, to read in world-wide newspapers, banner (and often fantasy) headlines following another experiment, attempting to involve the best minds in the world in trying to understand the nature of phenomena that seem trivial at a first glance, but which become incomprehensible under deeper inquiry – none of these are for the faint of heart.

Amicizia has been all of that, and even more.

I would recommend that the reader rejects stereotypes, and try to feel what lies behind all these tales, from Adamski to Sammaciccia, and of many others whose existence I am discovering only now, and who insist on anonymity. Contacteeism is a phenomenon much more widespread than one would suspect, and I believe that, aside from accepting or rejecting tales from persons who want to hide in the shadows, one must not forget the global scenario of the alien presence when considering stories told under the oath of the strictest secrecy.

In a scenario which has been evolving since Biblical times, we are confronted at different moments by leaders. They may perhaps be just shepherds, but sometimes researchers as well (i.e. Leonardo da Vinci – I suggest the reader to take a look at the hand of the angel that is present in the three versions of the "Vergine delle rocce"), and many more from De Fontenelle to Swedenborg, people outside their times, that are often forgotten by the history of sciences, a bit like the "Damned" of Charles Fort.

About the pictures in "Inside the Space Ships"

It seems that almost everything about George Adamski is highly controversial. I recently read a comment about the pictures which allegedly show Mr. Adamski peeping through the windows of a space ship. The commentator maintained that the pictures show some remote galaxy, with the images of windows and people superimposed over them. The idea is rather funny, so I decided to try to verify whether simple geometric calculus could provide not obviously, if not a definite answer, then at least a suspicion of plausibility for these pictures.

Let us suppose that the surface of the ship is rough. This means that an incident ray of light will not only be reflected according to the laws of specular optics, but will be refracted in all directions, with intensities which depend both on the direction of reflection and on the direction of incidence. A commonly used approximation is the following one. Let α be the direction of the incident ray, and β the direction along which we want to know the intensity. If we state:

$$\beta_* = \beta + \frac{1}{2} \cdot \left(\frac{\pi}{2} - \alpha\right) \cdot (1 - \cos 2 \cdot \beta),$$

we have:

$$\sigma(\beta) = \frac{1}{2} \cdot (1 - \cos 2\beta_*).$$

The function $\sigma(\beta)$ belongs conventionally to (0,1); it should be multiplied through a coefficient k<1, related to the physical characteristics of the surface.

If we substitute α by β in the previous formulae, we have the auto-reflectance function, σ_*, which gives the percentage of light reflected towards the observer.

The distribution of this function, for $0 \leq \beta \leq \pi$ is shown in the picture on the right.

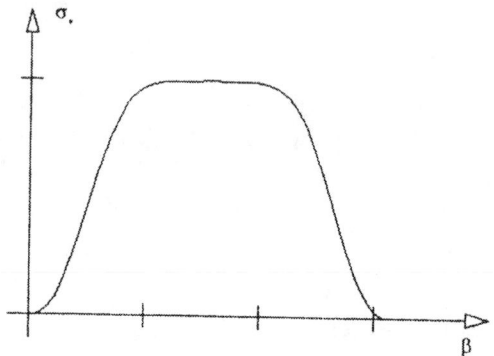

This means that a ray of light incident upon a smooth metallic surface will be mostly reflected, and, in some quantity, scattered around. For practical purposes, it is considered to be negligible the amount of light scattered beyond a limit angle τ, which is usually assumed to be very narrow (just a few degrees).

As a first trial to find out a possible geometrical environment for Adamski's account of his pictures, we might try the following. We refer mainly to the picture no. 14, opposite to page 151 of "Inside the Space Ships", Abelard Schumann, New York, 1955, which allegedly shows Adamski himself and another person peeping through the side portholes of a cylindrical space ship. The outer surface of the craft is lighted by a beam coming from an outside scout. Although Adamski's ships do not look to be exactly cylindrical in shape, we will start by assuming this to be generally their configuration.

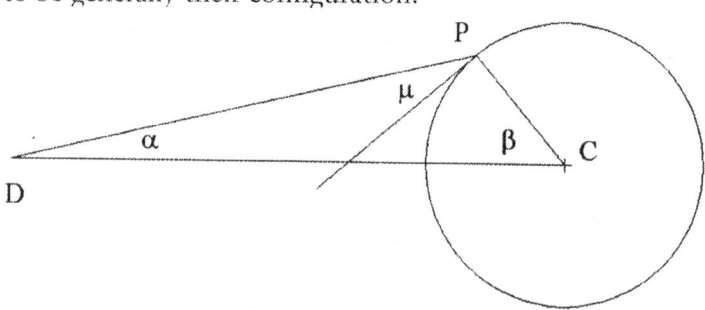

In the previous image, let C be the outline of the cylinder's axis, R its radius, D the external Scout (the source of light), d the distance between C and D, P the point where incoming light gets scattered at the limit angle μ, α and β the angles at D and C. We have:

$$\mu = \frac{1}{2} \cdot \pi - \tau$$

$$\frac{R}{\sin\alpha} = \frac{d}{\sin(\alpha + \beta)}$$

$$\mu = \frac{1}{2} \cdot \pi - (\alpha + \beta)$$

therefore:

$$\sin(\alpha + \beta) = \cos\mu$$

and:

$$\frac{d}{R} = \frac{\cos\mu}{\sin\alpha}.$$

The following picture shows the situation from above:

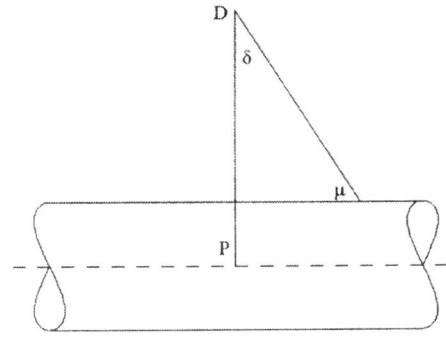

Clearly:

$$\cos\mu = \sin\delta,$$

and therefore:

$$\frac{d}{R} = \frac{\sin\delta}{\sin\alpha}.$$

Referring to the original picture in the book, let us assume that the lighted shape is an ellipse, and let us use a and b the lengths of its major and minor semi-axis. If f is the focal length of the camera, we have:

$$\tan\delta = \frac{a}{f}$$

$$\tan\alpha = \frac{b}{f}$$

whence:

$$\frac{d}{R} = \frac{a}{b} \cdot \sqrt{\frac{b^2 + f^2}{a^2 + f^2}}.$$

Now, obviously, we do not know the values of f, a and b. The only reasonable assumption we may state is that:

$$f > a$$

$$f > b$$

from which we may assume for d/R the limit of the previous expression when f is taken to infinity; we have therefore:

$$\frac{d}{R} = \frac{a}{b} = 6.15 \; ,$$

having assumed for a and b the linear measurements over the picture itself (namely 160 and 26 mm, respectively).

Assuming μ is 10°, we may therefore be able to compute the value of α:

$$\alpha = \arcsin \frac{\cos 10°}{6.15} = 9.21° \; ,$$

$$\sin \alpha = 0.1613$$

$$\cos \alpha = 0.9871$$

and therefore:

$$\sin(\alpha + \beta) = 0.9920$$

$$\beta = 70.79°$$

In order to try to find values for *d* and *R*, let's refer again to the photo, and let's measure the value of *h*, the height of one of the heads; we find h = 4 mm; if we assume that the real head is H = 0.23 m high, a scale factor σ arises:

$$\sigma = \frac{h}{H} = 0.016 \ .$$

In this image, let QJ represent the focal plane of the camera, H and N the bottom and top point of the head. It may be feasible to assume that point H, N and P are co-linear (the windows are 1.8 metres deep). Under this hypothesis, if x is the length of the segment PH, we have:

$$x = \frac{b}{\sigma} = \frac{0.013}{0.016} = 0.8125 \text{ m},$$

and now we may be able to estimate *R* and *d*. The following table gives different values of *R* and *d* (in metres), depending on different values of μ:

μ	R	d
86°	13.904	85.509
84°	9.274	57.035
82°	6.961	42.808
80°	5.574	34.279

From these figures, it results that there are reasonable values for τ (around 4°) which reflect plausible geometrical environments. A radius of about 14 metres for the ship is coherent with Adamski's tales about its interior, while some 90 metres of distance between the ship itself and the scout seem again feasible. Adamski himself (page 248) speaks of a distance of about 33 metres (100 feet), but this seems too low an estimate. The diameter of the scout itself should have been about 17 metres. These figures are found by examining the 2nd figure, and comparing its dimensions with what Adamski said on page 41 (total height between 15 and 20 feet), and on page 47 (cabin's diameter of 18 feet), and on page 246 (the scout was identical to the first one, to which these measurements refer).

Notes

1. Maldek is a planet, supposed to have exploded far ago, whose remains now constitute the asteroids belt between Mars and Jupiter.

2. I am unfair in saying so: in "The people of the Planet Clarion", pg. 51, Bethurum admits that this statement looks crazy, but he quotes some astronomers stating that it is possible!

Bibliography

George Adamski, Desmond Leslie: "Flying Saucers Have Landed" – The British Book Centre, New York, 1953;
George Adamski: "Inside the Space Ships" – Alebard-Schuman, New York, 1955;
George Adamski: "Flying Saucers Farewell" – Abelard-Schuman, New York, 1961;
Elias Ashmole: "Theatrum Chemicum Britannicum" – re-printed by Kessinger Publ. Co., Kila, 1991;
Colin Bennett: "Looking for Orthon" – Paraview Press, New York, 2001;
Jacques Bergier: "Les extra-terrestres dans l'Histoire" – J'ai lu, Paris, 1970;
Truman Bethurum: "Aboard a Flying Saucer" – DeVorss Publ., Los Angeles, 1954;
Truman Bethurum: "The People of the Planet Clarion" – Saucerian Books, Amherst, 1970;
Tullio Bosco: "Accadeva a Pescara" – Arte della stampa, Pescara, 1992;
Stefano Breccia: "Dalla Luna a Marte" – Achab, XII, 1990;
John W. Dean: "Flying Saucers and the Scriptures" – Vantage Press, New York, 1964;
Bernard Le Bovier de Fontenelle: "Trattenimenti su la pluralità de' mondi" – D'Anna Editrice, Firenze, 2002;
Daniel W. Fry: "Atoms, Galaxies and Understanding" – Understanding Publ. Co, El Monte, 1960;
Daniel W. Fry: "The White Sands Incident" – Best Books Inc., Louisville, 1966;
Bulat Galeyev: "Autour de Père Castel et du clavecin oculaire" – Colloquie International de Clermont-Ferrand, 1994;
Gavin Gibbons : "They Rode in Space Ships" - Neville Spearman, London, 1957 ;
Carol A. Honey: "Flying Saucers – 50 Years later" – privately published, 2002;
Howard Menger: "From Outer Space to You" – Saucerian Books – Clarksburg, 1959;

"Brother Philip": Secret of the Andes" – Neville Spearman, London, 1961;

Roberto Pinotti, Alfredo Lissoni: "Gli X-Files del nazifascismo" – Idea Libri, Rimini, 2001;

Wendelle C. Stevens: "George Adamski: Their Man on Earth" – privately published, Tucson, 1990;

Emanuele Swedenborg: "Le terre nel cielo stellato" – SeaR Edizioni, Parma, 1991;

George Hunt Williamson, Alfreb C. Bailey: "The Saucers Speak!" – New Age Publiching Co., Los Angeles, 1954;

George Hunt Williamson, John McCoy: "UFOs Confidential!" – privately published, 1958;

George Hunt Williamson: "Other Tongues, Other Flesh" – Neville Spearman, London, 1965;

George Hunt Williamson: "Road in the Sky" – Neville Spearman, London, 1969;

George Hunt Williamson: "Other Voices" – Abelard Products Inc., Scottsdale, 1995;

Lou Zinsstag, Timothy Good: "George Adamski – The Untold Story – Ceti Publ. Oxford, 1983.

The Ummo Affair

Beginning around 1965, hundreds if not thousands of physicists, engineers, biologists, astronomers, and selected people worldwide who were active in UFO study suddenly began to receive strange letters, written in many different languages, but primarily in Spanish, and originating from many number of places.

In some cases, the address was hand-written, imitating printed fonts:

Sr. Stefano Breccia

The authors presented themselves as extraterrestrials from a planet named Ummo, orbiting around a star, which they called Iumma, and which they equated (without being certain) with what our astronomers call Wolf 424 (in the constellation of *Virgo*).

The topics covered in the letters were extremely varied, ranging from the history of the civilization of the Ummites (in many instances, with often hilarious details!) to whole treatises on biology, mathematics, geometry and physics, with examples of applied technology, with comparisons between their way of life and ours, and between religions, politics, and philosophy.

In most cases, copies of the same letter were sent to several different addressees, every copy being individually typewritten. Presently the corpus of Ummite communications is really huge. Not counting multiple copies, we have some 10,000 pages (the description of their complicated language alone amounts to over 2,000 pages). As I have said, they were almost always written in Spanish, so that, in a way, Castilian is the official language of the "Ummo Affair". Words and even whole sentences in the Ummitan language are often intermixed with the Spanish text. Such interjections are written in capital letters. When quoting the Spanish text in the first Italian edition of this book, I left the original Spanish, thinking that Italians would understand it easily. However, I received so many complaints that I will now render them into English, although in doing so some detail will be lost.

Almost all the UMMO communications begin with the following header, which identifies the language being used (as if the recipient was not able to understand it by himself!), the number of copies sent, the date, and the address of the person it had been sent to. The word UMMOAELEWE identifies the central government of their nation (at the right the translation):

UMMOAELEWE
    Idioma: Español   Language: Spanish
    N.º de Copias 28   Number of copies 28
    Fecha 1972   Year 1972
    Sr. Stefano Breccia   Mr. Stefano Breccia

Let's begin reading (although not in full, because it is extremely long) the letter which, typically, was sent whenever a new contact was initiated. I repeat, the translation does not allow us to appreciate the peculiar way they were using Spanish. To maintain something of their stylistic tastes, I am going to include their mistakes (there are very few, I must admit) and will keep their strange use of quotes and capital letters.

Dear Sir,

We acknowledge the transcendence of that which we are about to tell you. It is clear to us that a statement of this kind should be considered preposterous, so that only a madman or a joker could have written it, or maybe someone looking for easy money.

When something new overcomes the usual patterns of public acceptance, without any means of verifying its reality, any balanced and intelligent mind has the right to and must adopt a skeptical attitude toward it. "Nobody has to accept a simple statement, especially in cases such as this, where its origin is unknown, and thus of a suspicious nature".

To us, what we are going to tell you is true, but we can't logically exect that you're going to believe such fantastic statements. "We must confess that, in your place, we would act the same way".

If YESTERDAY's concepts, which seemed fantastic and absurd had not been analyzed by competent scholars, would you have reached your present cultural level on Earth?

But, if such concepts are erroneous, if such tales are fraudulent, or if they are only the work of impostors, THE TRUTH has to be ascertained, thereby unmasking the deluded, the jokesters, the paranoiacs, or the impostors who tried to acquire the title of SCIENTISTS.

We have studied your history in enough detail to be aware of the immense bitterness caused by false prophets, those paranoid minds, who, in order to earn their own prestige, deceived the sincerity of Scientists acting in good Faith.

We have appreciated the struggle of your Science against Superstition, Astrology, Theosophy, Spiritualism and Dowsing, which, although they often carry in themselves a few principles worthy of being investigated, are at the same time full of unfounded statements, absurd methods of analogous reasoning, and of leaps of faith totally unacceptable to a balanced mind.

In these last years, because of the appearance in the Terrestrial Atmosphere of the so-called U.F.O. or O.V.N.I., vulgarly named Flying Saucers, the fantasies of men all over the Earth are overflowing thanks to the multiplication of fraudulent Newspapers publicity concerning such phenomena. Even worse, individuals such as George Adamski in 1952, Daniel Fry, or the Norwegian Edith Jacob in 1954 are appearing in various countries, pretending to have relationships with beings coming from other Worlds.

These subjects and the gullibility of the masses in immediately accepting at first impression their absurd tales have totally discredited the serious studies (os UFOs) that, in spite of everything, are being carried out on the part of the Official Technical Agencies of some Governments.

Then, as we are conscious that such tales have logically created a logical state of disbelief, it's not strange to us that it results in a repetition of the popular fable you call the Shepherd and the Wolf (Author: "Peter and the Wolf", or "The Boy Who Cried Wolf").

For such reasons, we are perfectly aware that we're not going to be believed, either now or in the future, whenever we choose to offer further revelations.

We repeat that our main goal is not rooted in being believed, without any more proof than these relations that we are going to send you. That doesn't mean that in the future we are not going to send you verified arguments and proofs of our identity, as we have already done in the United States, England, and Spain. Actually we are known in other countries. At this moment Australia, Germany, and the Soviet Union are receiving our reports, directed to prominent men of science, and though we recognize that a high percentage of the recipients tore our letters up, taking them realistically as the work of jokesters, or disturbed minds, in some cases the overwhelming collection of scientific data we offered was sufficiently convincing to many, that, in the end, we might be telling the truth.

Relations with these persons have continued, with them demonstrating reticence and discomfort, but objectively validating this strange situation, without totally discarding the hypothesis that our "Supposed" identity was likely true.

For this reason we are asking you to Read Carefully what we are going to tell you. It is Not important that you do so out of simple curiosity at first, disregarding totally the validity of our apparently fantastic statements. Again, we are not so ingenuous as to expect to be believed.

Moreover, we will give you a second reason that may seem paradoxical. We do not desire that the masses become aware of our existence. And we have very good reasons to justify this apparent contradiction.

Up to now, with very few exceptions, our contacts have been only with men of Science (True Researchers, and some Engineers), to whom we have offered certain data concerning Physics, Biology and Psychology, of interest to their work. At times even some technical procedures with industrial applications that, as in the case of A. W. R., from Atlanta (USA) have been accepted and patented. These were met with some surprise among experts knowledgeable in those fields, who assumed that it was the work of some whimsical and original scientist.

Then there are reasons that cause us to present ourselves to you, so that you may became part of this restricted group of people that, with more or less intellectual abilities, has become involved in the problem that laymen call the Origin of the U.F.O.

Physical data of our planet -

We come from a planet, whose name, Phonetically, could be written as UM-MO (the U very closed, almost guttural). The M could also be read as a B) whose major characteristics are as follows:

Orbit of "Ummo" -

Elliptical, with an eccentricity of 0.007833, with focus on the star we name IUMMA, or YUMMA, that performs the same role played by your SUN.

Distance from Earth to Iumma (Ummo's sun) -

The apparent distance, i.e. along the trajectory a coherent beam of waves follows in the three-dimensional space, was 14.437 Light years, on the fourth of January, 1955.

Real distance in the space with 10 dimensions -

This is a function of time. It must be computed in a Space with N dimensions, and shows some periodicity. It's very important to be able to evaluate it, because our Galactic trips depend on it.

Equatorial radius of "Ummo" -

Measured on the average surface of UAUAWEE. R=7251.61 kilometres.

Mass of Ummo -

$M = 1.36 \cdot 10^{24}$ *mass kilograms.*

Inclination of Ummo -

To the perpendicular to the orbit plane, it is 18° 39' 56".

Time of rotation around its axis -

600 UIW. Some 30.03 hours, equivalent to one day of yours, although a bit longer.

Gravity acceleration on Ummo -

11.88 meters/second (Author: this is a gross mistake, because accelerations should be given in m/s^2). *It may sound strange to you, but we aren't sure which star, in your catalogues, is our IUMMA (UMMO's sun). By means of a translation of coordinates, we estimate that from Earth our IUMMA can be seen as a star with the following characteristics:*

Right Ascension -

12 hours, 31' minutes, 14" seconds.

Declination -

9° hours, 18' minutes, 14" seconds (Constellation "Virgo").

Absolute visual magnitude -

"14.3"

Apparent visual magnitude -

It is reduced, because at 3.682 Parsecs distance from Earth there is a big cloud of cosmic dust, but should be between 12 and 13 (Author: another mistake: visual magnitudes increase in value when the amount of light decreases), so that you may see it only through photographic means.

Spectral type -

(According to Terrestrial Astronomical Classification) Type M. Our IUMMA is what you call a Dwarf Star. Unfortunately, the mistakes you've made in measuring distances are often greater than 15% among different stars catalogues you have issued. Therefore it's impossible to equate a star codified by us with one catalogued by Earth Astronomers, even with a careful translation of axes. In any case, we believe that our IUMMA may be the star which you catalogue as WOLF 424, as it's coordinates seem to be similar to the ones we have stated.

Geographical structure of UMMO -

Our geological-geographical structure is rather different from the one of Planet Earth. Oceans occupy 61.84%, with waters full of different saline chlorides. There is but one single continent, full of large lakes, the greatest of which, AUWOA SAAOA (Little sea of God) measures some $276 \cdot 10^3$ square kilometres. Our mountain ranges, very eroded, scarcely resemble Earth's mountains. The OAG OEII are a kind of volcanoes that look like big fissures that emit high and brilliant columns of incandescent mixtures of Methane-Pentane-Oxygen.

Language and characteristics -

We make use of a double language (by means of sequential repetitions of various terms we are able to express at the same time two different ideas). The words we use in this report are graphic expressions approximating their real phonetic sounds.

Mathematics of Ummo -

In mathematics we use a numbering system based on 12. (Of course we have only 10 digits, just like you). It's only for historical reasons. As a simple curiosity, we enclose here some examples of mathematical expressions, written together with their equivalent in our language.

$y = \sin 2\pi$		$\vec{A} \wedge \vec{B} = C$	
$\cos 2\pi$		$\sqrt[3]{27} = 3$	
$2 \cdot 4 \cdot 8 = 64$		ShU $\frac{1}{2}(C^v - C^{-v})$	
$\int Thxdx = \int nChx + c$		$y = \frac{dx}{dy}$	
$132 - 10 = 122$		$\Delta = \begin{vmatrix} 3 & 2 & 0 \\ 11 & 5 & 2 \\ 0 & 1 & 7 \end{vmatrix}$	
Tensor ϕ (Tenseur)		0 1 2 3 4 5 6 7 8 9 10 11 12	

Physical Units of Ummo -

Here we enclose some units utilized all over our planet.

Universal unit of distance of Ummo -

It is pronounced WAALI, and is equivalent to $12^{4.3}$ light years (Author: this means 43,700 light years, really a huge distance!).

Universal unit of longitude of Ummo -

ENMOO, equivalent to 1.8736658 meters.

Unit of time on Ummo -

UIW, equivalent to some 3.09 seconds, is defined as the half-life time of a mass of the isotope of thorium we call WAEELEWIWWOAT.

Frequency of the impulses of activation of nervous centres -

The frequency of the impulses of activation of nervous centres situated within the ventro-lateral coroidal plexus of the brain is $6 \cdot 12^3$ cycles/second (Author: 10,368 Hertz). *This is a frequency unit extensively used in Neurophysiology.*

Biogenetic constants -

$12^{-10} \cdot 6.58102$ seconds. It's the time required to integrate the quantic state in a carbon atom within the chain of Deoxyribonucleic Acid for the formation of one GEN.

Morphology of the Inhabitants of UMMO -

We inhabitants of UMMO possess a body whose physical morphology is analogous to that of Earth's Homo Sapiens. This is logical if you consider that the same biogenetical laws apply throughout the Universe. There are but a few small differences between you and us.

In a high percentage of the inhabitants of our Planet, the phonetic organ atrophies during the age of adolescence, so that the Ummite glottis suffers a sclerotic process which deprives us of verbal acoustical expression. Nevertheless, with the help of special prosthesis amplifiers of the weak frequencies emitted, we are able to speak normally, though the timbre of our voice is not as harmonious as yours, and the range of octaves is restricted.

Sociological characteristics of UMMO people -

Our species is much older than yours, and we have reached an elevated level of civilization. Our social structure differs considerably from yours.

Government of UMMO -

Our Government is ruled by the "UMMOAELEWE" (General Counsel of UMMO), composed of four members who have been selected among the whole UMMO species thanks to Psycho-physiological examinations. Laws are established as a function of the Sociometric Principles which prevail on our Planet.

Labour Coordination on UMMO -

Coordinating people's labour is attained by means of an efficient group discipline. Our economic structure is quite different from yours. Your curious institution of money is unknown to us, because all transactions involving the few valuable goods that exist on UMMO are accomplished throughout a Network of XANMOO (some kind of computers or electronic brains). On the other end, consumer goods (food and furniture) are of no value, because their abundance exceeds greatly the demand. Land and space are public.

Religion and creed on UMMO -

We are a profoundly religious society. We believe in WOA (God or Generator) and we have scientific arguments in favour of the existence of the BUAWAA (Soul). We are also aware of a third element in man that connects him to the A-dimensional Soul. It is lodged within the Encephalic Cortex, and we have informed two of your neurophysiologists of that. Our religion is so similar to Earth's Christianity that we were truly surprised when we discovered this. They differ only in the fact that our religion emerged in UMMO at a time in which our society was much more developed than yours at the times when JESUSCHRIST lived. Our kind of life, sexual habits, entertainment, and so on, are much more different than you could imagine. There are no true racial distinctions among UMMO's people, and varieties and biological species are less numerous than on Earth. We believe that this is due to the fact that the likehood of chromosome mutation is reduced, because our atmosphere is more protected than yours from the secondary effects of Cosmic Radiation.

Our first arrival on planet Earth -

Our first coming to the planet Earth happened largely because of the excellent Isodynamic conditions of space (space curvature) present at that time. Three of our OAWOOLEA UEWA OEM (that you call O.V.N.I. or flying saucers) landed in a place in the French Department of Basses Alpes, some 13 kilometres from DIGNE, and between 8 and 10 km from the locality of La Javie, at 4 hours and 17 minutes GMT (Greenwich time), on March 28, 1950.

We will provide details about the difficulties suffered by our first scouting group, which had to get through some very grave situations created by their absolute ignorance of Earth language and customs.

Other arrivals to Earth and dates -

Since then, more ships have arrived, culminating with the landing of three OAWOOLEA UEWA on February, 1966. The contacts with Earth's orography were: First, the arrival of two brothers of ours near Erevan (Soviet Territory) at 18 hours 47 minutes (Spanish time), in a place 6 kilometres *away from the Araks river on February the 6ᵗʰ, 1966. Second, arrival of two brothers at 20 hours, 2 minutes, in a place near Madrid (Spain), near the settlement of Aluche on February the 6ᵗʰ, 1966. Third, arrival of three brothers in a place near Townsville (Queensland, Australia) at 22 hours, 45 minutes on February the 6ᵗʰ, 1966 (Spanish time).*

Permissible Characteristics of our ships -

The characteristics of our OAWOOLEA UEWA, which are permitted to be revealed because they aren't subjected to our norms of security are: A) circular body, whose exterior diameter is 7,1 ENMOO (some 13.18 metres). B) Pseudo-lenticular transverse section 3,25 metres high. Provided with three extendable landing supports, with rectangular panels. The magnetic perturbations recorded by the Soviet engineer Alexei Krilov are not due to the magnetic criteria of propulsion. The strong induction he observed, in a point along the axis of the ship, ten metres away from its "Inertia Centre" often reached "600 Gauss", and was due to a secondary effect.

Aims and objective of the voyage -

Our aims in getting in touch with the Planet Earth are mainly centred upon Studying and Analyzing Earth Culture. It's difficult to state, in short paragraphs, the essential differences between our Civilizations, because, although our Scientific and Technological level is much higher than yours, we have to admit that you have cultivated some kinds of Arts (Painting, Sculpturing, especially Music) to a higher level than ours.

Unofficial Contacts with Governments and people -

There are strong reasons, which we have explained to some of our contactees, for us not wanting to be known on an Official basis, although Officials of some countries are now aware of our coming to Earth. Many Scientists and cultural Entities have been receiving our communications, all over Earth. But we must admit that results have been very poor, which was easy to foresee, taking into account the skepticism and good humour with which they have been received. When we brought revelatory knowledge, for instance about using tri-dimensional photography without any ocular visual media, or about interesting mathematic Developments, for instance concerning Graphs or Networks theory, they were considered as the work of some eccentric scientist, or a jokester. We do not tell you this with regret or bitterness, because from a psychological point of view such reactions are but normal. We wish to present you with a list of the Spanish persons we have contacted, letting them know our identity.

This letter is one among many instances, all similar to one another, of the Ummites introducing themselves. A typical mistake always present in this first letter consists of measuring accelerations in m/sec, instead than m/sec^2. However, in other contexts accelerations are measured correctly. Therefore we should infer that all of these first letters have been copied from a single original, which was mistaken on that point. There is another mistake in the first letter. When speaking about the apparent visual magnitude of Ummo, it should be, they state, something around the 14th magnitude. But, due to a big cloud of cosmic dust between Earth and Ummo, its magnitude decreases to 12 or 13. On the contrary, the scale of magnitudes is inversely logarithmic, so that a higher magnitude would apply to a dimmer light, and not vice-versa. Another inconsistency appears when they speak about the UIW. They say that a UIW equals 3.09 seconds, but, a few lines earlier, they stated that the duration of an Ummo day is of 600 UIW (30.03 hours). It seems, therefore, that one UIW equals some 180 seconds. Of course it is difficult to understand anything about the thorium isotope WAEELEWIWWOAT, but the easiest explanation would be that one UIW equals 3.09 prime minutes, not seconds!

Sometimes messages arrived, written in Ummite language, and therefore very hard to understand:

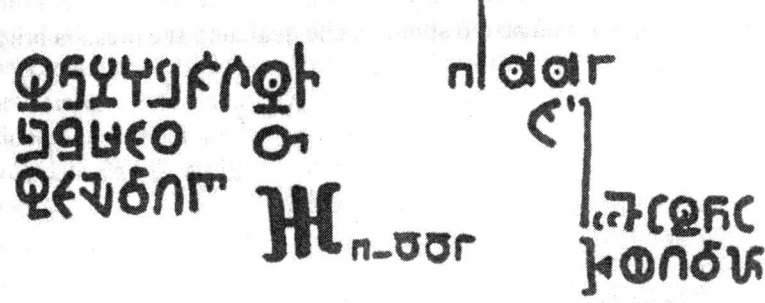

Nous vous présentons nos respectueux hommages

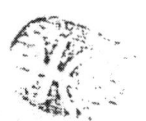

All communications end with the Ummo symbol as a signature. It was typically printed with a kind of stamp, which looks like the Russian letter Ж, or a "+" sign enclosed between a couple of reverse parentheses:)+(. The bending of the lateral bands may be more or less marked. At times, the horizontal line extends beyond the lateral one (as in the previous example). At times the symbol is rotated of 90 degrees (as in the following example).

Ummite letters were mailed from the most disparate places on Earth. The following message came to me from Leningrad (Saint Petersburg), as "Happy New Year" greetings:

Дек 1975

Уважаемый Господин Степан А. Бречйя

С Новым Годом из Умма!

These gentlemen addressed me in the classic Russian style, with first name, patronymic, and family name. I believe that very few people, mostly relatives of mine, are aware that my father's name was Aldo, whence the letter "A". Of course, some research at registry offices should have been enough to acquire this information, but it means, at least, that someone took the trouble to do so.

There is then a small mistake in the date (a "Г" is missing after 1975), but we can't expect too much!

How it began

Between the days of February 5 and 7, 1934, a Norwegian ship, off Newfoundland, made some wireless communications tests, using Morse code, at a frequency of 413.44 MHz. These signals in part went through the ionosphere, and 14 years later they were received on Ummo (Wolf 424, they maintain, is 14 light years away from Earth). Although the message was very short (just 6.8 minutes), it was evident that it was of artificial origin, and they were able to identify, in Galactic coordinates, the position of the star it had originated from. This star was baptized with the name GAA (= "square") because the image obtained from it reminded them of the analytical equation for a square [1]. Much later, it was found that only a Theodore T. Polk, from Pittsburgh (Pennsylvania) was aware of the meaning of this name. Unfortunately Mr. Polk passed away, without caring to reveal what he knew about it. Getting back to Ummo, the reception actually caused a commotion: it was the first proof of an offworld civilization. Therefore they decided to send an expedition to investigate.

They state that the space is ten-dimensional, and that the true distance between Earth and Ummo changes continually, because of the dynamics of such a structure. Only occasionally does it slip below three light years, and then it becomes within the range their ships. On February the 7th, 1949 [2], two ships reached our Solar System. The intense electro-magnetic emissions allowed them to identify at once our planet as the cradle of an advanced civilization. However they first paid a preliminary visit to Mars. About our "brother" planet they stated:

In this first examination, conducted from 290 KOAEE (25.926 km) away. One KOAE equals approximately 8.7 kilometre) an atmosphere was detected, not sufficiently thick as to allow the development of complex, multi-cellular beings. Many craters, and the crystal structure of its crust, verified by our instruments, showed no active bacteriological life, capable of decomposing into such a soil, at least in the place examined. Only later were we able to discover on this OYAA not only proteinic forms and amino acids, but also unicellular and pluri-cellular vegetal beings (it will not take too long for you to make the same findings [3]*).*

Indeed this last statement ("Ummo: La incredible Verdad", 1985) has not yet come true even today. Any way, once again, something is wrong with their computation: if one KOAE equals 8.7 km, then 290 KOAEE are 2,523 km, and their number (25.926) is wrong. Even if we read it as 25,926 the mistake originates in the evaluation of one KOAE. But if the KOAE is 87 km long, then the distance becomes 25,230 km, very close to their estimate. Of course, it's a rather long distance (roughly one tenth of the distance from our moon), but we may trust the accuracy of our friends' measuring instruments.

Leaving Mars behind, their scouts came towards our planet. Their exploration lasted a mere 4 hours (after so long a trip!), and took place 54 km high over Montreux, Switzerland. When they got back to Ummo, the recordings they had gathered were studied (a long description follows, which I will briefly summarize).

They had sighted aircrafts, roads, railroads (the latter's function not being understood). Because on Ummo houses are almost underground, they were not able to understand what Earth houses were for, and decided that they may have been some kind of factories. Some early mysteries were their discovery of pipe-like vertical structures inside the buildings, similar horizontal structures in vehicles, and small cylindrical objects that some humans were holding between their lips. Each of these items was seen to have a dark aerosol passing through them, which was identified as being composed of carbon oxides, hydrocarbons, and tar. They identified this aerosol as some kind of energy source. At first they believed the humans with these pipes (cigarettes) to be some kind of biological robots, that extracted the energy they needed from these tubes, but then they realized their mistake. However that posed another problem. They then decided that humans were not able to breath an atmosphere composed by nitrogen and oxygen, but had to add this strange aerosol to it. This might cause a problem for the members of a future landing expedition. Moreover it was not clear at all why some humans had cigarettes in their lips, and others had not.

Even worse than that, they observed persons walking without smoking, but at times lighting up a cigarette!

They identified our bi-sexual structure, but were not able to make a correlation between sex and clothing; in particular, the morphology of suits, and their chemical composition were not understood.

Other problems arose from the analysis of electro-magnetic emissions. They seemed to be modulations of acoustic signals (and they were horrified to discover so many languages being spoken on Earth!). Others were carrying binary signals (Morse and Telex), and the Ummites decided that humans had to resort to a universal binary language which they could not understand, in order to communicate over long distances. There were also large band signals, whose parameters were again a mystery. Perhaps it was another kind of language, but, strangely enough, such emissions (television signals) were present only in certain areas, and not in others.

They tried, without success, to ascertain whether our proteins were left or right-handed, and this uncertainty became another problem for their future expedition.

In any case, they decided to go on, and a landing expedition was organized. Six persons were to participate, two of which were women; they were expert in biology, structure of matter, communications, sociology and digestion pathology (the latter required because of their uncertainty regarding protein orientation).

The First Landing

On March the 28th, 1950, at 4:17:03 GMT, three ships landed in the area of Lower Provence, by the Cheval Blanc peak, along the Bleone river, at 2,322.95 meters above sea level (consider the absurd accuracy of such data, an accuracy always present in their letters). It must be noted that in the vicinity of Cheval Blanc only one place exists that satisfies the many descriptions of this event. Whoever wrote them, had to be perfectly aware of the orographic structure of the peak. The map I present in the Appendix shows this area. It is not clear at all why this peculiar place was selected for their landing.

Six persons were composing this expedition, four men and two women, mostly very young. They were:

OEDEE 95, son of OEDEE 91, the chief of the group: he was an expert on BAAYIODUII (biology), and was 31 years old;

UURIO 79, son of IYIA 55, expert on BIIEUIGUU (human psycho-biology), 18 years old;

INNDO 33, daughter of INNDO 29, expert in DOLGAA GOO (physics of the structure of matter), she too was 18;

ODDIOA 1, son of ADAA 65, expert on AYUU WADDOSIA (communications), 78 years old;

ADAA 66, son of ADAA 65, a technician on AYUYISAA (sociology), 22 years old; he has been the only one to die, because of an accident, on November the 6th, 1967, in Yugoslavia. His body could not be recovered;

UORII 19, daughter of OBAA 7, expert in the Pathology of the digestive system, 32 years old.

It may be noted that Ummites are identified by a name composed of an alphabetical portion and a number, followed by a patronymic, whose meaning is not clear. More over in some cases the names of fathers and sons differ. On this subject, they have stated:

> *... we get a qualification that results into some coefficients, or indexes, which constitute our UMMOGAIAO DA (identifier). You may interpret that as a kind of a digital analytical imprint which reflects our physiologic-mental personality ...*

What could have been the mood of this little group scouting our planet? At first, they dug an underground refuge (the "Galería"), and started looking around, remaining concealed inside it. In the distance they were able to see the town of Digne, with its old cathedral, conspicuous among the small houses, whose function was not clear to the Ummites; 8 km away lay the village of La Javie, plus small farmhouses here and there. Obviously the place in which the Ummites concealed themselves was totally secluded.

On the evening of the 29th they decided to start exploring the vicinity; two of them went out on foot, strongly armed, wearing protective suits. Between two high trees the two found some burnt rocks, which a quick chemical analysis identified as limestone. Some meters away they discovered something extremely interesting:

> *Some 1.8 ENMOO (one ENMOO equals 1.9 m) they saw fragments of yellow-white plates; they were irregular and flexible, crumpled, and filled with signs, or characters, evidently written by humans; Three among them were soiled with faecal remains. Many unknown flying animals flew away. This discovery was judged so important that they went immediately back to the Galería ...*

The microscopic structure of those sheets of paper was analyzed, and found unknown, because on Ummo they do not use cellulose to manufacture paper. Characters were found to be not hand-written, but impressed by means of some kind of standard punches. Evidently a liquid ink had been used. But what was puzzling our friends was another question:

> *The presence of faecal remains was an enigma. Analyzing the excrements, epithelial cells were found, surely coming from human intestinal glands. A list of possible hypotheses was drawn up. The explanation that looked most probable involved some ritual: perhaps, at times, humans, when they did not agree with what was written on a sheet of paper, were accustomed to soil it with faecal remains …*

It looks that this "find" is still today kept on Ummo as a kind of relic, in thermostatic conditions. Later on, our scouts found out that they were the first pages of "Le Figaro" of 26-27 of March.

The following day, they saw eight vertebrate animals, some 350 meters away. These had sharp appendages above their heads. Two men were sent out to investigate, wearing a peculiar kind of dark protective suit, that made them look like friars. Heavily armed, and very cautious, the two men emerged from the Galería. On Ummo there are no cows, and therefore those animals were unknown to them. They were able to identify their sex, thanks to their udders. At 15 meters from the beasts, the Ummites stopped, to record sounds and images, and to activate

> *… the process of detecting electro-static and gravitational fields generated by those animals, that, although having noted the presence of our brothers, went on without any reaction.*

All of a sudden, a young shepherd appeared from within the nearby rocks; his apparel was different from that which was recorded over Montreux. The two Ummites weren't able at first to ascertain his sex. They were very worried, remained motionless, and asked their chief "via wireless" what to do.

The boy, looking at them without too much surprise, had in his hands an yellow-whitish substance, coveted with something black, and he was quietly eating it in front of our brothers.

The boy uttered some sentences that could not be understood, then approached, lifting one hand to cover his eyes from the sun. The Ummites took this as a kind of greeting, and did the same, puzzling the boy. After a few seconds, when nothing happened, the shepherd turned, gathered his cows, and went away. During the encounter, the two Ummites had stayed still, trying to answer the supposed greeting, according to what their chief had told them to do. They returned to the Galería, very troubled. Their surprise had been so sudden that they had made no recording of the boy, and therefore now there was nothing to study. They had no idea on how things could turn out, so they strengthened their shelter, waiting for whatever, perhaps even an assault by the human military, if it had been notified by the boy.

The third day passed, without any human assault. Thanks to their optical instruments, the Ummites again sighted the boy, still engaged with his cows, and a couple of men further away. This time they recorded everything, so that our friends were able to manufacture suits similar to the ones the men had been wearing. Buttons, whose function had not been understood, were made with an aluminium alloy, neckties with a kind of flexible stuff (they had not been able to understand the topology of the tie), and an handkerchief peeking out of the breast-pocket was imitated using some alimentary substance. Of course, the "suits" were identical to one another.

The 2nd of April, as the situation was still quiet, three Ummites put the suits on, and emerged, again heavily armed, looking for the cow-herd. With them they were carrying a brotherhood message, which had been prepared on Ummo in a binary language easy to be understood (!), and a sheet of the newspaper they had found days before.

The boy greeted them, lifting one hand. The Ummites, not knowing how to behave, remained still. The French boy approached, and said something. Then, he unwrapped some of his food, which he presented the Ummites. One of the three, well aware of the risk he was taking (remembering that they had not been able to ascertain the orientation of proteins), decided to accept, warning his companions to abstain. He ate, and survived. So the hypothesis that all over the universe proteins are oriented in the same way was verified. In the meantime they were perplexed because the boy wasn't reacting to a language never heard before. The Ummites had understood that on Earth many different languages are spoken, but they believed that humans were able to understand them all. Later on they found that the boy's name was Pierre, he was 11, and that a couple of years before he had encountered two German topographers, and was therefore a little accustomed to non French-speaking people. It has not been possible to identify him, from the Ummites' descriptions.

As there were no unfriendly reactions, the Ummites started at once to learn from the boy. Gesturing, they had him spell the names of near-by objects, and to read (albeit with some difficulty) passages from the newspaper. From the Galería their friends were following the event, anxiously. By nightfall, the Ummites had learned 110 French terms; but the binary brotherhood message had not been understood by the boy!

After having gained a certain control over the environment, the Ummites got bolder, daring a night excursion into the village. Then, with a true "coup de main", they invaded a near-by farmhouse, after having put its inhabitants into an hypnotic trance, stealing many different objects in order to study them, and collecting biological specimens from the people inside. When they recall this episode, the Ummites seem to ask forgiveness. They say that later they found a way to compensate the people for the stolen items.

The head of the local *Gendarmerie* recalls that one morning old Violat had come to report the theft of the most varied and often undesirable objects, but still adding up to the value of 70,000 francs, a huge sum in those times. Some years later, Violat having passed away, his sons won a lot of money in the lottery, moved to the *Côte d'Azur*, and were never heard of again …

Soon, the Ummites had realized that they could easily acquire a mock identity, pretending to be Dutch or Swedish physicians or biologists, looking to start a scientific and commercial enterprise in that area. In this way, the first Ummite beach-head on Earth was established.

The Severed Hand

After this first contact, the Ummites established a stable nucleus in France, beginning to perform biological research on our environment, and financing themselves through various cover activities. What you are going to read is the report about a regrettable event, which took place in this period (someone has suggested that this history marks the beginning of HIV, and this hypothesis is not so strange as it may seem). Both the human and the Ummite versions of this story exist. The latter, however, arrived some fifteen years after the event. The Spanish press called this episode "El caso de la mano cortada", for reasons that will soon be made clear.

The chief character of this history was Doña Margarita Ruiz de Lihory, marquise of Villasante, and baroness of Alcahali, owner of many villas and estates throughout Spain. She had studied law, and was a rather good painter. Her first husband, the American Richard Shelly, had passed away in the forties, leaving Doña Margarita with their four children, three boys and a girl: Juan, José, Luis and Margot. Afterward, she lived with Don José Bassols Iglesias, a Catalan gentleman, who some pretend to have been her second husband.

In spite of her aristocratic background, Doña Margarita was not very wealthy; but, one year before the even of this story took place, her situation had improved. Some Dutch physicians had leased, at a very high price, the "Cuarto del Moro", the cellars of her house in Albacete. Moreover, some lottery wins, and an inheritance from an unknown American relative had improved her financial state.

On January 19th, 1954, Margot died in the Albacete house, of a pulmonary oedema. Her body was transferred to Madrid, and buried two days later. On the morning of the 20th, with her daughter's body still in the funeral chamber in Madrid, Doña Margarita and Don Bassols went to claim another lottery prize, 100,000 pesetas, a huge amount of money in those days. On January the 30th, Luis, one of Margot's brothers, denounced his mother to the Madrid police, stating that she had mutilated her daughter's body. After having exhumed the body, it was found that in fact one of the hands had been severed, and that part of the tongue had also been cut off. Worse yet, both eyes were missing.

In denouncing his mother, Luis had added unusual details: the woman, 67 at the time, was extremely fond of animals, and was surrounded by cats, dogs, birds, and other animals. When one of them had died, she usually performed a post-mortem examination, recovering some of its organs (head, tongue, hearth). This was confirmed by Don Jaime Aguado Trigueros, the veterinarian who was in charge of the animals. A search of the house in Madrid yelded a plastic jar, filled with alcohol, containing an amputated hand in it, shreds filled with blood, scissors and other surgical instruments, two skinless dogs' heads, viscera and other animal remains. The hand had been amputated with masterly skill, and was the missing one.

Confronted with the evidence, Doña Margarita declared that her daughter's hand was a kind of holy relic to her, and that she worshipped it as though it were a saint's relic. This action not being strictly illegal, the accusations were dropped.

During next year Doña Margarita, Don Bassols, their domestic staff, and many friends, all passed away, in a series of different accidents. Only an American was able to escape, taking with him some microfilms of unknown content, then vanished.

That's the conventional account. Fifteen years later, the Ummites acknowledged being heavily involved in the affair. At the time they had firmly settled in France but, for some unknown reason, they had decided to engage in covert biological research in Spain, passing themselves off as physicians. A first attempt to buy a zootechnical farm had failed, and so they had the idea of opening a veterinary clinic in Segovia (remember that most of them were experts in biology). They therefore sought the help of someone from the locality. They researched in the archives of the Deuxième Bureau in France, and discovered the figure of Doña Margarita Ruiz de Lihory, who had been active in the resistance during World War II.

Among the things they discovered about Doña Margarita, were her studies in psychopathology and her love to animals. Two of the Ummites settled therefore in Albacete, and managed to get in touch with the noblewoman, apparently by chance. Then they became friends with her, pretending to be physicians from Northern Europe, and suggesting to her the way to recover from a chronic headache. Having won her confidence in this way, they asked to lease from her the large cellars of the house she owned in Calle Mayor, which she rented to them at a very high rate (the Ummites had understood the economic importance of gold, and were discreetly selling some here and there, discretely, in order to raise money. The rent was therefore not a problem to them). They asked only for the strictest secrecy about their presence, and actually only Don José Bassols Iglesias and a maid were aware of two "Dutch physicians" having leased the cellars.

At that time the most important activity of the "Dutchmen" was the analysis of immune-globulins in vertebrates, of electrophoretic velocity and of the corresponding evaluation of the amino acid sequences in the polypeptide chains of diverse antibodies. In particular, they wanted to verify whether mutations could be induced by the stimulation of what our biologists call the "variable portion" of the concatenated network of globulins, in a way similar to what occurs on Ummo.

They had imported a virus from Ummo, unknown on Earth, similar to the one we name "rat polioma virus", but provided with a much more complex capsule. The Ummites verified that this virus was not able to multiply in healthy tissues, but that it could perforate the mucoproteic layer of the cellular membrane in the case of a necrosis in the cellular area, and then to begin a quick process of reproduction. While commenting on the events in Albacete, the Ummites say they have means which permit them to detect the situation of these and other viral colonies, when the number of individual viruses is in excess of $420 \cdot 10^6$ individuals per AXEESII (the unit of volume for biological evaluations, that is equal to 36.77 cubic millimetres), equivalent to 11,693,818.24 individuals/mm^3, measured at a distance not greater that 78 meters (Author: The usual absurd precision of Ummite data.

By the way, the original text contains here another mistake: instead of $420 \cdot 10^6$, they write a simple "12", which is obviously preposterous).

On September 4, 1953, during routine remote control inspections, 18 centres of infection were found on the bodies of Doña Margarita, her daughter, a neighbour, and on the bodies of three dogs. Evidently, something had gone wrong. These infections were destroyed at once, without the knowledge of the affected people, but, to prevent further troubles, four more Ummites were summoned from Austria, with more sensitive instruments. Using these, 26 more infections were found within Albacete, and they too were destroyed. But the risk was still present in concentrations below the sensitivity threshold. These were usually harmless, but ready to erupt in presence of a local necrosis.

Time went by. Some minor infections were found here and there at times, and, when possible, were destroyed without the knowledge of the affected persons. At other times infections were left untouched, because they would have required a surgical operation, too difficult to accomplish under these circumstances. The area around Albacete was scrutinized on a regular basis. At last, someone realized that Doña Margarita was the owner of a certain number of houses throughout Spain, and that since the beginning of this incident, she had already made a couple of trips to her house in Madrid. On November the 7th they went to Madrid, and found out that Doña Margarita's daughter was affected by six centres of infection, all of them very deep, and therefore impossible to destroy remotely (Author: This statement seems a bit strange, because the regions interested, according to our friends, were the ocular globes, the epithelial tissue of the tongue, and the skin of the palm, structures that seem decidedly superficial).

Probably due to its contact with the Earthly environment, the virus had undergone some mutational change in its structure. The situation was desperate. The Ummites (always under their covert identity) disclosed the situation to Doña Margarita and, at a certain moment, were to get in touch with the Spanish medical authorities, to inform them about these ongoing cases. Then, in the accounts they have sent us, they stated that it was decided to do nothing, out of the fear of not being believed. Most of the Ummites present in Europe gathered in Madrid, trying to find a solution to this problem, but without success. A surgical operation was impossible without revealing the situation, and the top leader of the expedition decided that no action was to be taken.

On January the 12th the Ummite biologists gathered in Madrid decided to wait for the girl's death, and then to perform a biopsy on her body to understand what mutations the virus had gone through. A few days later, Doña Margarita's daughter passed away. Her hand, ocular globes and tongue were amputated, with the goal of understanding what had gone wrong. Her mother was silenced with a large amount of money, made to appear as a lottery prize.

The Ummites have not thought it fit to tell us the results of their investigations, aside from stating that *"... seen through the eyes of an uninformed human of Earth, such things would be judged as astonishing, and repulsive."*, which is likely to be just a euphemism.

It is interesting to note that just one year later, the Italian researcher Renato Dulbecco was to begin his studies on the rat polioma virus, which were to win him the Nobel Prize in Medicine in 1975.

Their Physical Characteristics

The Ummites, as it has already been told, declare themselves to be almost identical to humans in their physical structure, and that this peculiarity allows them to live in our environment, without being noticed. Of course, the laryngial atrophy presents some problems, but probably the members of the expedition had been altered to avoid this.

A strange problem arises with their fingertips, which, according to them, are very sensitive. They maintain that they are not able to use a typewriter. All of their letters have been transcribed by human typists. When they were forced to write by themselves, the Ummites used an handwriting, which simulated typing.

Another unusual characteristics of theirs involves their perspiration, which seems to generate some unexplained problems with nearby dogs. For this reason, the Ummites began to use deodorants or scents in order to prevent such problems.

Moreover, they are morbidly sensitive to being seen naked. For instance, when one of them has to undergo a surgical operation, he/she wears over his/her skin a transparent and inert sheath, which does not prevent their surgeon from seeing the body, but which quiets their phobia. In a hierarchic society such as the Ummite, one of the worst punishments given to a person guilty of some offence is to force him/her to appear naked in public! This practice seems to date back to very ancient times, and the following text may give us a bit of an explanation:

> *On Ummo the ordinary greetings consists of placing one's hand over the breast of his/her friend; but, without any doubt, we are still chained to past habits. For instance, although it is no longer mandatory, when a brother wants to express his/her good will to a superior, he/she puts his/her hands over his/her thigh. Of course, this sign recalls*

ancient degrading attitudes that goes back to the most remote Ummite times.

In the times of the higher OGIIA (the highest chiefs on UMMO) the OEMII (a generic man/woman) had to express his/her subjection by placing his/her hands over the genitals of his/her superior, and a prerogative of the latter (both sexes) was to sexually abuse their subjects. The role of servicing the superior is not considered incompatible with the self-respect of the OEMII. This is a rather strange custom, which Earthmen may not understand, because they have been educated in a different way.

Amazingly, when the Ummites are suddenly accelerated upward or backward they experience … an orgasm! The following cartoon, drown by the scholar Jean-Pierre Petit, and dedicated to Ribera, illustrates this unusual situation:

Their society is of course very different from ours, but one would not expect such a strange difference.

In the Ummite "Middle Ages" there was a climate of savagery and sadism. During this period, vivisection was commonly performed, both for research, and as punishment. It seems that IE 456, daughter of NA 312, was the most despotic tyrant in Ummitan history. The following quotations are taken from "Ummo informa a la Tierra" ("Ummo Reports to Earth"):

> *When 8.5 years old (according to Earthly reckoning), IE 456 was a young girl characterized by an astonishingly high intelligence quotient, and the Local Council (in those days parents had no authority about the instruction of their children) sent her to the ONAWO UII (something like a University, or a Polytechnic Centre) to study OOLGA WAAM (Physics and Cosmology). At 13.2 years of age IE 456 was appointed teacher of WAAM-TOA (History of Cosmology), and began her researches into Ummo's gravitational field. Her NOA (pupils) were all older than her. That little girl was not very healthy, but she used to impose a rigid and wicked discipline upon them ... When the spirit of the old chief OES 17, son of OES 14, dematerialized (in other words, died) the young IE 456 was elected to the council of AASE OUIA (governors) ... despite the general consensus being immediately against this decision ...*
>
> *17 days after her nomination, the famous AAR GOA, (the girl) provoked what you call a "coup d'état" ... and proclaimed herself the highest authority of Ummo ... Then IE 456 declared that God is but the sum of all Ummites, and that she herself – IE 456, daughter of NA 312 - was going to be WOA's (God's) brain! Today it is difficult to believe that such a young girl, who had not even experienced her first menstruation, could have subjugated millions of people in this way.*

When 15.2 years old, IE 456 promulgated the famous twelve INAIE DUIO (laws or decrees), decreeing her as the WAAMDISAIAYA (Central Coordinator of the Cosmos):

She proclaimed herself as the owner of all beings over Ummo, and that she might dispose of their life as she alone saw fit;

She designed and standardized the worship rituals that would henceforth take place all over Ummo in her honour;

She declares that the ultimate purpose of Ummo is Science, and, should it be necessary, that the whole population has to die on behalf of scientific necessity, she would no scruple to prevent it from happening.

A period of true terror ensued ... women, men, boys and girls who were not able to pass the standard levels at the BIEEWIA (psycho-technical tests) were to suffer the destiny that, on your Earth, is reserved to guinea pigs. For instance, in AEVO UI ONAAWO (AEVO's University), in a single day, and in no longer than 12 UIW (about 36 minutes), were 160 young people vivisected, of both sexes, aged between 17.6 and 22 years, excising, without anaesthesia, part of their encephalon, in order to look for the olfactory centres within the cortex.

The mythomania and the raving cruelty of that young girl ... were generating disgraceful excesses. Only the highest scientists, scholars, philosophers were allowed into her presence, and they had to go before her totally naked, at a time when sexual modesty was very strong among our ancestors. Their eyelids were closed with some glue.

. . . . .

For the first (and actually the last) time in Ummo's history, IE 456 imposed her daughter, WIE 1, daughter of OOWA 33, as her successor to the government of Ummo ... WIE 1 began to exert her power over her terrified subjects at an age even younger than her mother's, on the very day the latter had passed away. She was only 12.2 years old, and this time there was no protest against what was to be called UMMOTAEEDA (which could be translated as "infantocracy", that is, government by babies).

But WIE 1 did not have as brilliant a mind as her mother. On the contrary, she revealed only an ordinary mediocrity, but nevertheless she outdid her mother in sadism, idolizing her body in a paroxysm of egolatry and narcissism. It is in this historical period that the sublime figure of UMMOWOA appears (whom we will describe later), thanks to whom people started to return their thoughts to the true God. For the first time a religious worship was held. A bomb, exploded perhaps by one of her female servants in revenge for having been treated cruelly, ends the life of WIE 1 (as cruelly as her mother's). Our history records more than four million victims during her short government.

And we do refer to Nazism! The UMMOWOA whom they mention corresponds strangely to the Jesus of Christian tradition, and his life marked a change in the planet's history, although, by their own account, Ummites do not seem very different now (maybe they lack just vivisections). Continuing with quotes:

The joy with which the news of WIE 1's death is received was followed by a reaction of hatred against everything that smacked of Science. The wonderful labs USADAADAU were burnt during a single night ... Luckily, no casualties were recorded or rather, just one, innocent: Pure Science, whose evolution was interrupted for many long years. On the contrary, research in the fields of Philosophy, Telepathy and Mathematics sprang up again, like a beautiful column of burning methane (Author: on Ummo the nights are embellished with columns of volcanic fire).

Ummitan society itself underwent a profound transformation. The terrible experience they had suffered with autocracy forced a restructuring of the Government. Polycracy appeared, in the form of the UMMOAELEWE (General Council of Ummo), which has remained until today, although with some minor changes.

By the way, the so-called Symbol of Ummo is, to be precise, the symbol of the UMMOAELEWE. Let's get back to UMMOWOA:

Compared to Earth, we are older people, which started to become organized when the various dispersed groups on our planet associated with one another, and gave birth to a monocratic government. In later times, many OGIAA (top chiefs) succeeded, and imposed rather dictatorial laws ... But the infamous AAR GOA (illegal seizure of power) by IE 456, daughter of NA 312, in 1301 gave birth to the haunting crisis we have already described. During this period, in 1282/3 of the Second Time, under the supreme Chieftainship of that good and paternal ancient OGIAA OES 17, son of OES 14, and in the flourishing industrial district of IOSAAXII, the Divine UMMOWOA was born.

The Industrial Plan developed by IE 456 when OES 17 passed away had reached all the sectors of society. Millions of men and women were forcefully exiled from the colonies where they had been born, to work as slaves or to be utilized in biological experiments. Among thousands, UMMOWOA was integrated as a slave worker in the construction of IUMMASNEII (a solar energy plant), in the SIUU plain ... (Author: Details are given about the life of IE 456 and WIE 1). *Such things happened in the*

Ummo year 1368 of the Second Time, but in those days UMMOWOA had already started to spread his teachings. At night, when the slave workers were returning from the OYISAA DOAA (a kind of encampment), after having stripped (It's an obsession! - Author) *and being searched to prevent the theft of copper, that strongly-muscled young boy of sweet countenance spoke softly to hundreds of people, who were listening in silence ... His listeners were not only from humble, uncultured classes. Among them technicians, physicians, biologists, philosophers, professors that the despotic regime had condemned, depriving them from their professions. The man was galvanizing all of his listeners with the purity of his logic, with the objectivity and humility of his words ... When his guardians, moved by his sublime doctrines, offered him a preferential treatment, at their own risk, he declined. He was developing himself as the incarnation of WOA into an OEMII (living body), and his power of persuasion was such that none, among the intellectuals that were following him, could object to anything.*

. . . . .

(A long analysis follows, with comments of modern psychologists about the recordings of UMMOWOA).

. . . . .

His exhortations were disseminated clandestinely all over Ummo. Writings, DOROO (photo-acoustic tapes), oral reports faithfully transmitted by children, and above all telepathic transmissions to relatives and distant friends, despite the strict vigilance by the Police, reached the most remote places on our planet.

Everybody was aware of the existence of UMMOWOA, but the tacit silence that had encircled him for many years preserved the secret about his identity. But the pleas of his most faithful followers were useless. When UMMOWOA finished his REVELATION mission, he announced that he would deliver himself to the authorities, who had been looking for him without success. On 15 of the year 1402, at night, with the cunning complicity of the Police, the divine UMMOWOA fled from the SIUU plain, in order not to compromise his strictest followers, and presented himself to the WOODO (police) of the town AASE GAARAADUI (the ruins of which are still preserved).

Thus began his haunting martyrdom, which was described to us by exceptional witnesses of the time, along with newsletters and official papers wich have been analyzed critically by our experts ... UMMOWOA was included within the sorrowfully celebrated GROUP of persons selected to be victims of scientific experiments, and under the explicit order of WIE 1, requiring that physiologists determine procedures that would subject the victim to the most painful death. His martyrdom took place in 1405.

. . . . .

The specialists involved in his vivisection were astonished when the body of UMMOWOA disappeared in front of them at the moment of his death. On the YOAXAA (a kind of surgical table) remained only the GIAA DAII (sheets and porous cloths), but his extracted viscera, his blood which had soaked the UBOO (plastic sponges), together with his cephalic liquid kept in a receptacle, all vanished at the same time.

Despite the efforts of the police, the news spread istantaneously. The biologists were called by WIE 1 to render a detailed account of what had happened; this account excluded any possibility of a collective hallucination. The girl listened to their testimony with a superstitious fear, *but instead she accepted the version presented by other "scientists", who denounced the biologists as impostors, so that these latter were put to death ...*

A very detailed analysis follows of the report made by the biologists who had been conducting the vivisection, with comments from which the atrociousness of those times emerges. Then an examination follows of the parallels between UMMOWOA and Jesus.

General Features

In Ummite society there's nothing similar to our entrepreneurship; they maintain that ... *all Ummo people could be considered what you call state employees*. Nor any free enterprise is possible among them. Their central government takes charge of youngsters from very early age: *Boys and girls start their education when 13.68 years old, in something like a UNIVERSITY, which resemble small towns and are kept under strict discipline. The whole of their future life depends upon the results of this training period. These centres are called UNAWO UE.* When a boy (about 15.5 years old) selects a girl as his possible wife, he must *submit his choice to his teachers, who will search, through the network of SANMOO AYUBAA (brains, or computers of Ummo), for any possible physiological or mental incompatibility which might jeopardize a future marriage. If no problem arises, the boy reveals his intentions (kept secret up to this point) to the girl.*

Of course, the plan of studies and subsequent placement into the society are managed directly by the state, using the all-pervading computer network, and continuous testing during the process. Their society doesn't use money, because of the abundance of goods available to meet all requirements. In any case, they have a somewhat similar method of exchange: the access to goods is regulated (as usual!) by the government, through a very complex mathematical formula, that takes into account the capacity of the individual, his/her cultural level, level of responsibility, and efficiency. According to this formula, a higher hierarchical level doesn't automatically imply a better "wage". The opposite happens rather frequently, when a low-level person who does his/her best, and achieves a high efficiency rating may "earn" more than his superior, if the superior doesn't have similar proficiency. The equation is too long and complex to be quoted here. In any case a lower efficiency level is given more consideration the higher the level of the person involved. In theory such a mechanism should lead to an extremely efficient managerial class, for efficiency (in the sense of term used in the physical sciences: "costs versus results") is the main parameter that determines one's capacity for acquiring goods.

In a strict sense, there is no a private ownership, of course, but there is a *de facto* property. For instance, a house is assigned by the state to the new couple (there are no "single" people), and it is theirs to use as long as the couple is intact. When one of the two passes away however, the surviving one leaves it, and enters collective dwellings, or goes to live with one of the children (again, it's up to the state to decide what must be done). Every benefit is assigned by the state in usufruct, on the basis of the above mentioned formula, and may be taken back at any moment. Among us the money is the mechanism that regulates the access to goods, but among them this function is usurped by the state.

Aside from this continuous overshadowing of the state, their society is strongly male-oriented, with the wife subjected to her husband. In the workplace there is no distinction between sexes, but within the family the husband has a heading role. As I have said, in the hierarchy and at work there is a substantial parity. As an example of that, here is the tale told by one of the typists from Madrid who was employed by our friends.

It seems that for unspecified reasons, the chief of Ummite expedition had to spend one night in Madrid. This person was to arrive to Madrid from Singapore via London. They asked the typist if he was willing to host their chief; after consulting with his wife, the man accepted. At night three Ummites arrived at his house, two girls and an older man. The "chief" happened to be the younger girl, YU 1, daughter of AIN 368. The other girl, UUOO something, was her secretary, and the man was a high level person in the hierarchy, but obviously of lower rank than the girl.

After having secured the house and the street, with other Ummites standing guard in the neighbourhood, they began their supper (boiled potatoes, hard-boiled eggs, and fruit – no wine) with only the young chief actually eating, while the other two stood behind her. While dining, YU 1 explained that UUOO was a mathematician, but that, after a mistake she had made somewhere in Mexico, she had been downgraded to the role of her secretary!

When the supper was over, YU 1 decided to help the host's wife with doing the dishes. Then, although having been offered the bed of her hosts, the young girl insisted on sleeping on the dining room floor, again with the other two keeping watch. At sunrise the three Ummites left, but before that the two Spaniards observed what appeared to have been a reproach that YU 1 made to her assistant – for although the chief had spoken in Ummitan, it was evident that the other girl blushed, and her eyes filled with tears. Before leaving, YU 1 told the housewife that it was unpleasant to see that in Spain women were less accustomed than males to read, and ordered the other girl to give her an encyclopaedic book, in Spanish, as a reward for her role as host.

The typist has obviously paid utmost attention to what took place inside his house. He tells how when YU 1 was asking her companions something, they lowered their eyes before answering, and that they never took the initiative to speak directly to their chief. Another example of how rigid is the hierarchy inside their society!

Returning to Ummo, their houses are of course rather complex. Here is a plan:

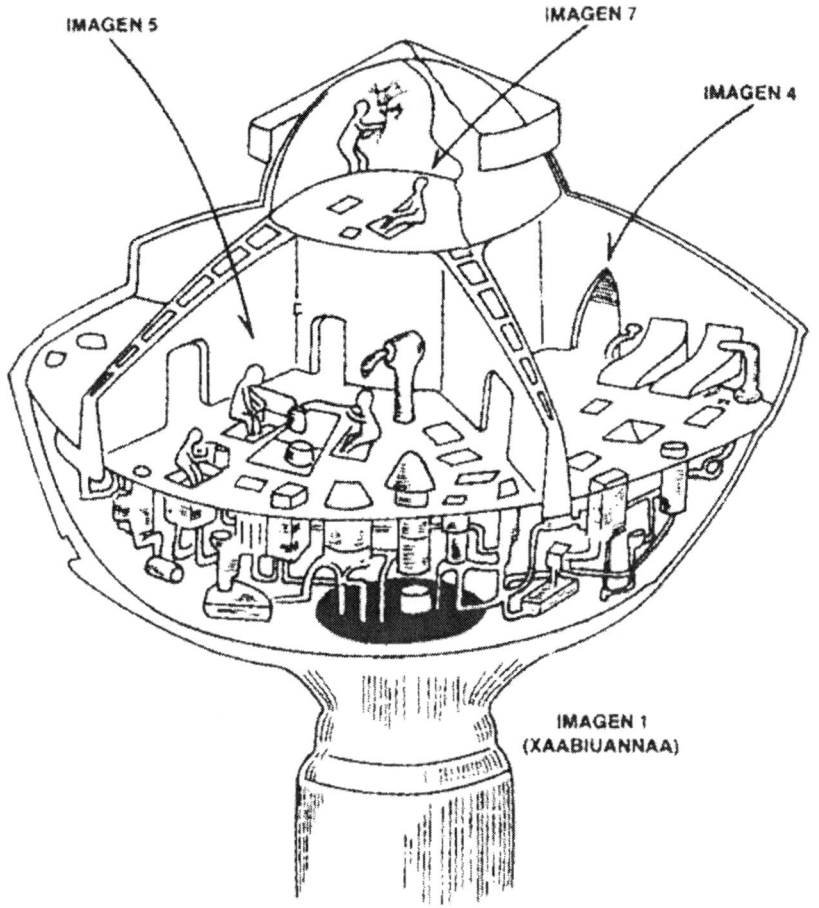

This structure is usually more or less underground, but it may be raised on demand, thanks to a piston that may be seen at the base of the vertical rod, which is pushed upward by the conversion of a mass of sodium from solid to gaseous. As usual, it would take too long to examine all of the details our friends told us about their homes. Generally speaking, under the main floor there are the control devices, that receive supplies from outside, using a kind of pneumatic dispatch, recycle wastes, generate energy, and so on. Technology is very complex, to the point at which what looks similar to one of our fork is for them an electronic gadget that would require a degree in engineering to be used by a human. Among the many unusual characteristics of these houses is the capability of modifying the use of each room. A bedroom may be changed into a kitchen, and so on. Things are not as easy as they may appear. For instance, the room where breakfast is served resembles the Japanese one, with persons sitting on the floor, their legs dangling over a hole warmed – in Japan – by hot coals. In addition, thanks to the piston, the house may rotate about itself, if told to do so.

The OAWOOLEA UEWA OEMM

This is the name Ummites use for their space-ships, whose structure is at the same time elementary, but very complex in its details. There are many reports by the Ummites on this subject. Let's start with a section:

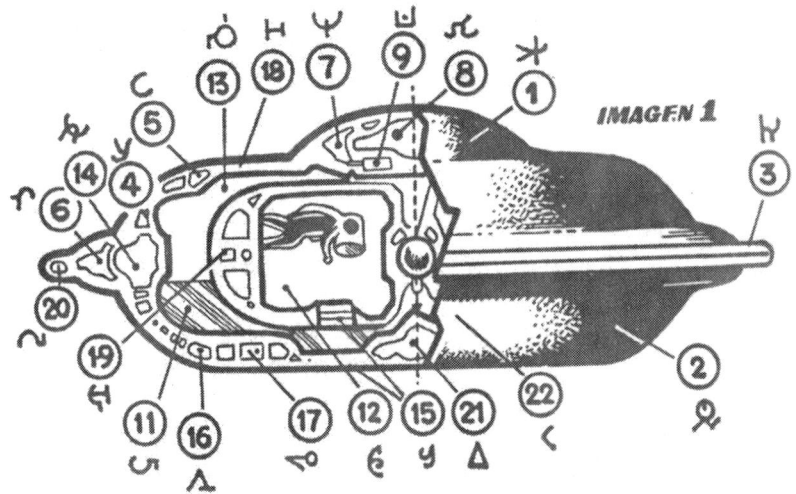

IMAGEN 1

Referring to the numbers in the figure, this is the list:

1. ENNOI: Prominence, tower or dome, in the upper hemisphere of the UEWA OEMM (its surface is transparent)
2. ENNAEOI: Central body of the ship's upper structure
3. DUII: Ring, or equatorial rim, that goes around the UEWA
4. AAXOO XAIUU AYII: The toroidal generator of magnetic fields
5. NUUYAA: Toroidal store of oxygenated water and melted lithium
6. IDUUWII AYII: The propulsive device, distributed inside a ring-shaped space within the DUII
7. Energy generator. It transforms lithium and bismuth into energy, through a plasma state
8. IBOZOOAIDAA: Central equipment that controls the inversion of IBOZOO UU

9 XANMOO: Peripheral computers (the central computer is situated within the geometrical centre of 12 – AYIYAA OAYUU, the central sphere)

10 Censored

11 TAXEE: Jelly mass, that often fills the inside of the AAYIYAA OAYUU

12 AAYIAA OAYUU: floating cabin

13 YAAXAIUU: It may be translated as "Magnetic cavity"

14 Within this toroidal structure many of the UEWA's devices may be found; part of the magnetic field generator, control devices for XOODINAA, food stores, equipments for manufacturing parts, and so on

15 IMMAA: Some of the access hatches

16 YAA OOXEE (MERCURY SUPPLY)

17 A complex ring-shaped hull, that contains, among other things, retractable landing gear, elements transmutation devices, and so on

18 XOODINAA: Membrane, skin, external wall, or protective armour of the UEWA; it is opaque, (not in ENNOI, which is transparent), and very complex in its structure

19 (text missing)

20 YUUXIIO: Toroidal equipment to control gaseous environment

21 UAXOO AAXOO: Armoured emission and detection centre

22 ENNOI AGIOA: Assembly dome or cone, which may be dissolved or regenerated under control of the central XANMOO

It must be noted that the central cabin (12) floats inside the craft, within the liquid (11).

The shape of the craft, they maintain, is a compromise among various possibilities. The best rotational shape is the one on the far right:

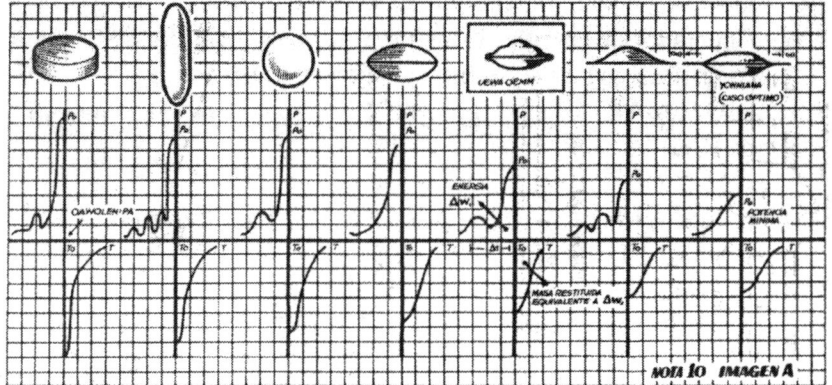

It corresponds to the function:

$$z = H \cdot e^{-k \cdot R^2}.$$

In another report, we find a curious mistake, where they reproduce the following formula (in x and y), obviously a mistaken one:

$$y = \frac{1}{\frac{1}{e^{x^2}}}.$$

The cabin (12) is toroid shaped, and filled with a gelatinous substance (DAXEE) whose purpose is to protect passengers from the very intense accelerations (up to 245 m/sec²) that take place in fractions of a second. The thixotropic principle is in use, that is to say, the capability of some substances to immediately change their state, from solid to liquid through external stress (for instance impacts or vibrations) and to return instantly to their previous state. This is, for instance, the cause of quicksand, a compound of sand and water, which is usually solid, but when an unfortunate victim walks on it, the pressure generated by his weight is sufficient to transform at once the substance from solid to liquid, so that the fellow sinks into it. When the descent stops, everything reverts to a solid state! Also, the thixotropic properties of some kinds of mud are used when drilling oil wells. The ship from Ummo is filled with a thixotropic liquid. When the ship accelerates, or slows down, everything becomes solid, protecting the occupants in this way.

Obviously, to survive within the liquid, the occupants must wear a kind of space-suit, at least to be able to breathe. Actually the space-suit functions go beyond that. The description of just the helmet (the object resembling a cylindrical box around the head of the Ummite which can be seen in the previous image) is more than ten pages long.

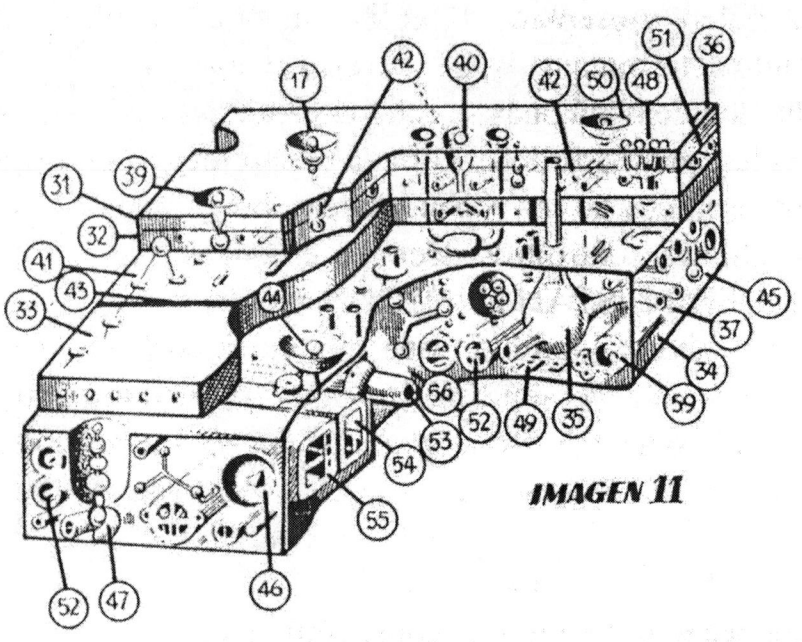

IMAGEN 11

The structure of the external skull is also very complex in itself, full of micro-pipes and microscopic devices. Thixantropy is widely used to cope with minor external damage, such as the impact of small particles. Other microscopic tubes generate a coating of liquid lithium, which refrigerates the external surface. Others have optical fibres inside, and various sensors to monitor environmental conditions. It is really a truly complex structure, but a plausible one providing that one has the technology required to make it.

The Ummites state that each data channel within the ship is based on three independent lines, each one working on a different physical principle, so that three malfunctions at the same time are extremely improbable. The principles used are: laser signalling, gravitational waves, and a third one unknown on Earth, which they call nuclear resonance. This is what they say about the latter:

For instance, think of a set of a few atoms of Molybdenum, let's say Mo_1, Mo_2, Mo_3, ..., Mo_n, whose nuclei share the characteristic that at any moment their configuration of energetic levels is the same, vis-a-vis the distribution of nucleons. It does not matter if the quantic levels of their electronic shells are different, nor that their orbitals are engaged in some chemical connection. We say that these atoms OAWOOENII (ARE IN RESONANCE WITH EACH OTHER).

We know, by the way, that whatever an atomic particle (a neutrino, a proton, a K meson, and so on) is actually a projection into the three-dimensional lattice of the same mathematical entity, that we call IBOZOO UU (up to the point that, within our WAAM (Universe) we attribute the characteristic of EXISTENCE or REALITY only to the IBOZOO UU).

Now, suppose that, within the atom Mo_1, we may change the orientation of a single NUCLEON (for instance a PROTON). It may happen that the inversion is not absolute, in which case you could see the conversion of its MASS into ENERGY:

$$\Delta E = m \cdot C^2 + K ,$$

m being the mass of the PROTON, and K a constant. This way we get an isotope of niobium (as you name this fundamental chemical element). But we may force the disorientation of the "axes" of the IBOZOO UU in such a way that, to an observer, the PROTON would disappear, without any emission of ENERGY. You might believe that such a phenomenon contradicts the universal principle of the conservation of mass and energy (conservation, by the way, correctly doubted by some physicists on EARTH. Actually the hypothesis of some EARTH scholars about the CREATION of MATTER within the UNIVERSE is based upon the fact that indeed sets of the IBOZOO UU invert to our three-dimensional lattice, being in this way perceived by ourselves, who live within it).

The remaining text is full of mathematical expressions, where human and Ummite symbols are mixed together, so that they are almost impossible to be reproduced via a PC, so let's stop here. In summary, they say that they have a chain process, with the creation of a niobium atom corresponding to the energy loss of the Molybdenum atoms, a process that decreases with the distance. Its interesting property consists in that the speed of its propagation is <u>infinite</u>.

Continuing with the description of their transmission systems, there is an obvious incongruity in that two out of three (optical fibres and gravitational waves) have a finite time of propagation, while the third one propagates instantaneously. Their modulation is done according to what we call the Fourier Transform, that is the analysis of frequency distribution, in a way very similar to our Pulse Code Modulation, with a packet frequency given by:

$$N = \Delta F \cdot \log_{12}\left(1 + \frac{S}{R}\right),$$

where ΔF is the channel bandwidth, S and R the intensities of the signal and the noise (*Ruido* in Spanish).

Some Other Techno-Scientific Information

In the following, we are going to have a look at selections from the enormous amount of information available in the Ummite reports. Of course, it is not feasible to condense in a few pages such huge a body of data and therefore this description will be sketchy.

Let's start with the theory of parallel universes (which was later been presented by the Nobel laureate Andreji Sakharov – it is not clear whether he too had been receiving letters from the Ummites):

> *Today we are aware that there is not just a single Cosmos (our own), but an infinity of couples of universes. There is a duality inherent to cosmology. The differences between two elements A and B of each couple consists in that their electrical charges are opposite to each other (You erroneously refer to this phenomenon as Matter and anti-matter). For instance, out twin universe does exist, but:*
>
> - *in its atoms, their shell is composed by positive orbiting protons (positrons), and their nucleus by anti-protons;*
> - *the two universes will never be able to get in touch, and it makes no sense to believe that they may, one day, be superimposed on one another, because they are not separated by a dimensional relationship (that is, it's meaningless to say that they are distant a certain number of light years, or to assert that they coexist in the same time);*
> - *both twin cosmoses have the same mass, and the same radius, corresponding to a hyper-sphere with negative curvature;*
> - *but the two universes present distinct singularities (in other words, in our twin Universes there are not the same number of galaxies, nor do they have the same structure). Therefore there is not a twin Ummo, nor a twin Earth, as one might naïvely believe. This is not just a hypothesis. Later on we will explain this;*

- *both cosmoses have been created at the very same time, as we will explain later, but the arrows of their times are not oriented in the same direction. That is to say, it is illogical to say that our twin universe coexists with ours, or that it existed before, or will exist later. The only thing one may rightly assert is that it EXISTS, but not now, before, or later. On the other side, its evolution must be equal, and parallel, to ours. As an example, let's suppose that a man, on a planet of our twin universe, is able to live eternally. If he starts his clock when his universe is born, and stops it when his universe collapses, the time he notes will be the same as in our universe.*

We could say the same about the infinite couples of universes that exist within the WAAMWAAM (Pluricosmos). Let us remember that the Pluricosmos doesn't look like a universe. In the latter, galaxies are like islands floating within an immense sea. But this sea is actually a sphere of many dimensions, and at least we may speak of the distance between two galaxies, and also of the gas in between. It is much more difficult to envisage the WAAMWAAM, because the couples of universes exist in the Nothing. It's useless to think of distances, or even that such distances are null; such an image is deceptive. There is more, which shook our scientists, when they discovered it: our twin cosmos exerts an influence upon ours, although no space-time relations exist between them. We have been able to perceive the existence of our twin cosmos thanks to such influences, in the same way that you, while watching TV, perceive a car passing by from the interferences that appear on the screen. Of course, our universe has the same influence upon the other one, but the asymmetry of these influences implies that our twin cosmos has a different distribution of galaxies.

Interesting, isn't it? By the way, to bypass another long quote, I will clarify here that if the name of our cosmos is WAAM, the name of our twin is UWAAM.

Changing the subject abruptly, let us discuss about titanium memories:

> In Earth's digital computers, components called arithmetical units using transistor circuits perform elementary operations with great speed. UMMO gets the same effect using IYOAEE BOO, which are based upon chemico-nuclear reactions, on a very small scale, instead of transistors. To do so, we use a few hundreds of such basic reactions, designed so that computations are performed in the base 12. For instance let's look at this sum, and its subsequent verification:
>
> $$12+1=13$$
>
> The result of the reaction (in which are involved not billions, but just a few micro-masses, perfectly controlled).
>
> $$C_6^{12} + H_1^1 = N_7^{13}$$
>
> The result of the reaction is analyzed with great precision, following with a sequential repetition. Over a block of titanium three rays, of infinitesimal section, are incident, at a very high frequency (therefore they are able to get across the block), without affecting the nuclei, but only the shells of atoms; the frequency used is around $8.35 \cdot 10^{21}$ Hz, and are different for each ray.
>
> Such high frequencies are outside the characteristic spectrum of TITANIUM, therefore these rays, one at a time, are unable to excite shell electrons. But this is no longer true when the three rays get, all together, upon a specific ATOM.
>
> The interaction of the three different frequencies generates an effect that you have known for a long time, and call it

> BEATING or HETERODYNE, whose result is a lower frequency, that coincides with one of the spectral lines of TITANIUM.

> Therefore the atom becomes excited, and, the three rays being capable of orienting themselves with great precision, are able to reach any atom they are looking for. The decoding process (during which the involved atoms are taken back to their original state) is obtained in the same way.

There follows a rather interesting long discussion on cameras (both for still frames and for movies), too long to be quoted here in detail. We are going to take a look only at the unusual structure of their lenses:

> We are able to exert a very detailed temperature control within a solid, liquid or gaseous mass. By emitting two rays of very high frequency, we are able to modify the temperature gradient at a point (P) within a GASEOUS mass, that is, to warm a small space surrounding it. Making use of a correct range of wavelengths, we are able, within a gaseous mass, to create an artificial environment, where some places are at a different temperature than others.

In short, they are able to create gaseous lenses. Comparing our cameras to theirs, they are astonished by the complexity of one of our zoom objectives using crystal lenses, feeling their approach to be much simpler. Leaving aside their very long description, let us have a look to one of their cameras:

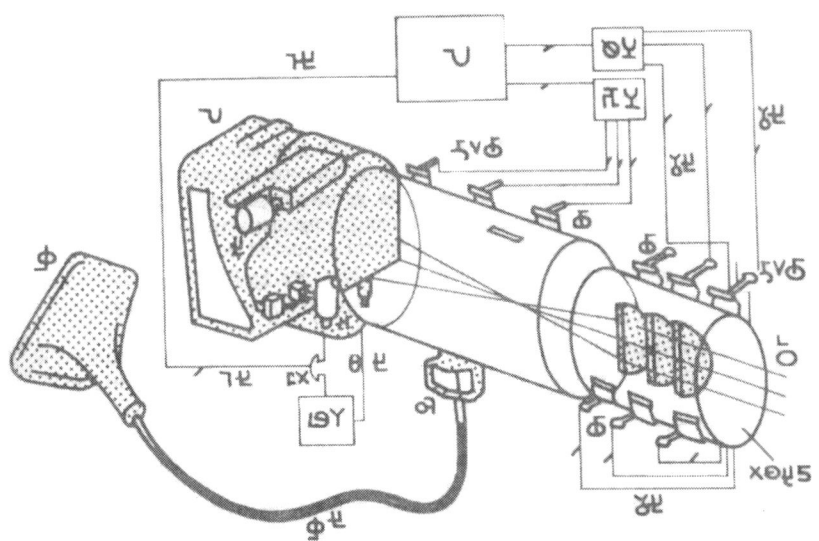

In all appearances, it is a device based upon "gaseous lenses", plus electronic deflection of light, and of course minus photographic film. Images are stored (in 3 dimensions), within the usual titanium memories.

The Landing at San José de Valderas

In the late afternoon of the 1st of June, 1967, many Madrilenians were enjoying the last rays of the sun in the meadows around San José de Valderas, in the area of Aluche, a few kilometres away from Madrid. All of a sudden, they were astonished by the appearance of a large disc-shaped object, with a transparent dome on the top of it, revealing the Ummo symbol on its lower surface. The object was emitting an orange light, howered for a while at mid-air, then shot away at terrific speed. It seemed to land near to a restaurant (*La Ponderosa*), then started again, and was quickly lost in the distance.

Because of the relatively long duration of the accident, two different persons (one of whom has never been identified) were able to shoot pictures of the object. The next morning, Antonio San Antonio, a reporter from *Informaciones*, a Madrid evening newspaper, received a telephone call from an unknown individual, who refused to identify himself. The man told the journalist that, the evening before, he had taken pictures of a "thing", that had appeared in the sky all of a sudden, and that he had personally developed the film during the night, and that he had left the prints for the reporter's disposal at a photo shop.

San Antonio went to the shop, and found an envelope waiting for him, with five pictures inside, together with their negatives. Strangely enough, the negatives were not arranged as a single strip of film, but where cut into individual items, as if their maker had decided to keep some of them for himself. This supposition is confirmed by the fact that the negatives bore the numbers 12, 19, 21, 23 and 24, so that the photographer must have kept at least 8 other pictures. Moreover, the photographer was evidently an amateur, because the pictures had not been developed by the best process. That same day, *Informaciones* published a couple of the pictures. The five photos were to be numbered Y1, Y2, … , Y5.

The second photographer, that evening, had been one Antonio Pardo (Only the person's name is known, and as it is rather common in Spain, it was not possible to get in touch with him later on). Pardo too had developed his film by himself during the night, but he kept his pictures for more than two months.

A couple of months later, Márius Lleget authored a book ("Mito y realidad de los platillos voladores" = "Myth and reality of the flying saucers"), and, at the end, he asked that anyone who had witnessed any strange phenomena to contact with him. That's what Pardo did.

In a very long letter to Lleget, Pardo told him that he had taken pictures of the object, that he had developed them by himself, but also that he had made inquiries in the area where the UFO had appeared. In this way he had found that the owner, and some customers of the *La Ponderosa* restaurant had witnessed the short landing. During his inquiries, Pardo had spoken with a boy, who told him he had seen nothing, but that, the next day, he had found a quantity of little metallic tubes. Using pliers, he had been able to cut open one of them. He said there was some water inside, that immediately evaporated, and a rolled plastic strip. Pardo bought the tube from the boy.

Later on, many more small "tubes" were to be found:

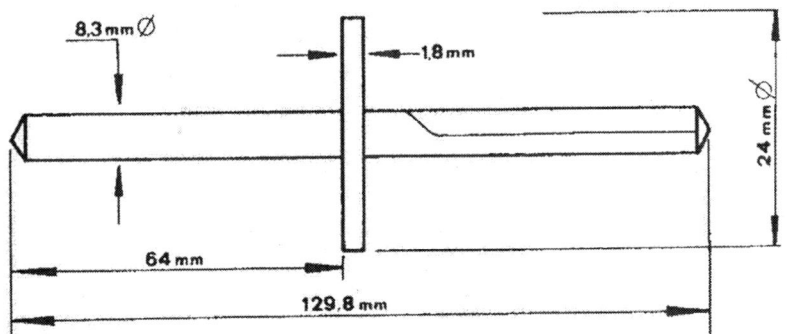

Then Pardo sent two negatives to Lleget (they were to be named X1 and X2), of the nine he claimed to have shot. By having seven pictures at his disposal, and by examining their background, it was possible to find out where they had been shot. The X photos were taken from a single location, while the author of the Y pictures (that Pardo claims to have seen), had moved around a bit. Thanks to this examination, it was possible to reconstruct the trajectory of the object:

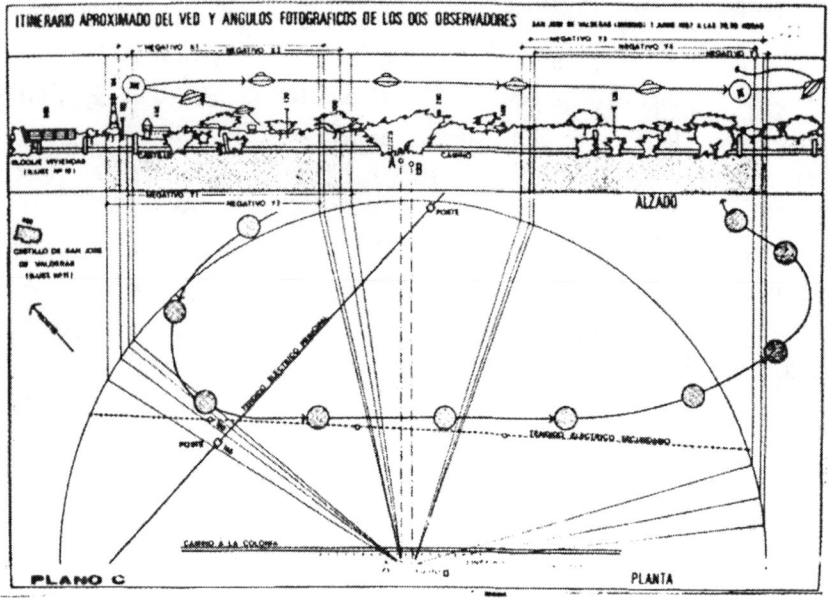

Examining the two sets of negatives, it was easily found that they had been shot by two different cameras, both in the same format 24 x 36 mm. format.

It is interesting to note that, one month before, on the 20th of May, the newspaper *Informaciones*, printed in Alicante, had published a letter, received by a prof. Sesma from Madrid (one of the many who were receiving reports from the Ummites), in which he predicted the coming of one of their ships within a end of the month, and that the ship was to land somewhere near Madrid!

The usual humorous treatment of the incident followed: many among the witnesses (who had not seen the pictures) told newsmen that the symbol beneath the UFO was a dark "M", which obviously had to stand for "Martians"!

Over all, some fifty witnesses to this landing have been identified, some military among them. At the landing site three rectangular marks

(15 x 30 cm) were found, 12 cm deep; these indicate a weigh of around 15 tons for the object.

Many small tubes were found on the ground. All of them had one or two plastic strips inside. These strips measured 2 x 13 cm, and on each one there was impressed the Ummo symbol.

Strangely enough, a certain Henri Dagousset sent a letter to some shops in that area, offering 18,000 pesetas (at that time, some $ 300) for each cylinder supplied to him, and 7,000 pesetas to the shop-manager who would act as intermediary. The letter gave a "poste restante" address in Madrid, which turned out to be non-existent.

One of the strips, and a piece of a cylinder were chemically analyzed at the Air Ministry, Instituto Nacional Técnica Aerospacial. The results prooved interesting enough.

The metal was nickel, of high purity. The plastic strip was found to consist of a peculiar kind of vinyl poly-fluoride, a substance which (at that time) was produced only in the States, and supplied to NASA! In those years, that particular substance was of no practical use, so it was nowhere to be found. NASA, its buyer, was using it as an external protection for some parts of rocket motors exposed to the open air before being launched.

As it may be seen from the diagram, the measures do not correspond to any metric system (the "tubes" were all equal to each other). The plastic strips were very stiff, therefore impressing the Ummo symbol upon them must have been a long process, using a very hard punch, and a powerful press. Of course all of this doesn't prove an extraterrestrial origin for the cylinders, but it does show that an hypothetical forger would not have had an easy time making them.

The "gurus" of today's UFOlogy tend to consider the pictures as fakes, because the clouds in the background are different from one another! Since the sighting lasted some fifteen minutes, this should not present a problem, but that's the way things are.

The Encoded Message

In October, 1974, the Ummites sent an urgent message to their Spanish friends, stating that they were estimating as highly probable a war between the United States and the Soviet Union, with a probability of around the 28% (to-day we may be happy that they were wrong). Had the situation deteriorated, the Spanish people could have gone to seek refuge inside an bomb shelter that the Ummites would kindly place at their disposal. (Antonio Ribera, the late well known UFO scholar, underlined that, to reach the shelter by following the Ummite instructions, would have taken some 20 hours, so that they would have arrived there when the war was likely already over!). To the letter an encoded message was enclosed which, according to them, contained the instructions to open the shelter. Should such a necessity arise, Spaniards were to receive specifications on how to decrypt it (which, according to the letter, would have required two hours, using paper and pencil).

Here is the text of the message (dashes correspond to words who, nobody knows why, are not encoded):

AFIRTMFIOEMPRNGHHYSMROEMMIDJLORMOINTO----ENUIMVBIORMNURMOUD
ENVQITMNVOIYMNDHFYRMBIUPMDNYRMCIZAEEITNNEWWYSIITRFGGHOPWBVN
ENCVURNORMORMNCTURMBCNYTRDNCVIKFHISMNAEUNRTEDANERSITROCHROS
NTRIMTRUERTSSWERE
EWRIENMTYUDMURMNVBFTYRGDKORMNAYUNRETYENCVIOMDRETYENURSERTYU
QUEMZBORYANCXRIRPKSEWQNYRMAKDENVZMOPLŇERTSJYEŇAQQEYRNNIENVD
REYUAHGDRIRTYUNCVQUYYEBCUOSHYRNNNERSNWUNBOPEPPPERANXQIRQQQ
AREUNCUEEORPSRRREBC----------------ERTUDNNZXWIDQARAERAERA
DRUNNIODNIRNIENMPLKAGEEKISUSIOADUNDFFERAWWOXEOPAFTRURUEHRTY
RETYDUENOLARTENORGDAAQYENBBVDETYRNZSIENORMNAYERTAUNCYEIRMAR
ERTYSGOFTEBAIW------ERIENKDFEUMIOANSERTESERTERTENXDREYUIOSD
ERTQUNSOEYANXOEXXUENPRTDDRENVESIRNDREIMCTRUDAOQPEEWNCFDYEIM
QYENSDERTCMPRJLKAMŇDEEWNLLEŇWQYEŇEWŇEWIEHDNNEIRTELAEROERQIM
ERT----------------------EROERMERTEORE ERTEIERTEJDFEROI
UENQWOSHHDHWLDRWTEMVBERIOPLASRETYDNETRYUE---------------
ERTYUENCBVORTYENVBTRUEOERTANCBTERYUSFDREMBCNUERT
ERTEGDFERTEURTYERADSFGDRETURO----------QTERETRUETREIERTENCB
ERTEREFSORTQOASSETESSSERETBCVNMPOLAREFAIEROOFHUQWNVCFRETFDG
TERESDFRETY.
-RETEFDHFSRETYSDRET-VCTYERTEFDRTEHDFERT
YRRTREURETYERTFFDGTRFYHDFREAMŇŇOERŇOQWEBFDRE--

Lines are 59 characters long, and 59 is a prime number. One can't but feel that the first three or four characters of each line are a key to the decoding.

The frequency distribution of the letters is that of the Spanish language; therefore the coding does not affect this distribution (the letter W is present, that does not appear in the Spanish alphabet, but it's a simple assumption that one character corresponds to a blank, so that it tallies).

While trying to decode this text, a problem was immediately: the Spanish language does not have a standard collating sequence within the alphabet. For instance, some place the ñ after the n, while others place it after all other letters. Therefore, every key-based algorithm must make attempts with different alphabets, and that, obviously, takes time. Any way, as far as I know, nobody has been able, up to now, to decode the text.

This is the frequency of the letters inside the message:

e	15.044	q	2.163
r	13.373	w	2.065
n	7.571	b	1.967
t	7.375	v	1.868
o	4.818	h	1.672
d	4.818	p	1.672
I	4.523	g	1.180
y	4.425	l	1.082
u	4.425	ñ	0.885
m	4.031	x	0.787
a	3.736	k	0.787
s	3.540	z	0.492
f	3.048	j	0.393
c	2.262		

and this is the frequency in conventional Spanish:

_	15.614	m	1.871
e	10.936	b	1.228
a	10.175	g	1.111
o	7.778	v	0.994
s	7.135	y	0.936
l	5.965	q	0.643
r	5.731	j	0.526
n	5.614	h	0.409
i	5.205	f	0.351
d	3.977	z	0.292
t	3.684	ñ	0.175
u	3.509	x	0.058
c	3.450	k	0.058
p	2.573		

The first character, "_", stands for a blank; trying to match the frequencies and to substitute each letter with the corresponding one in the sequence, does not lead to satisfactory result:

TCREODCRS_DQEAJYYNUDES_DDRLKHSEDSRAOS----_AIRDVGRSEDAIEDSIL
_AVMRODAVSRNDALYCNEDGRIQDLANEDPRXT__ROAA_BBNURROECJJYSQBGVA
_APVIEASEDSEDAPOIEDGPANOELAPVRÑCYRUDAT_IAEO_LTA_EUROESPYESU
AOERDOEI_EOUUB_E_
_BER_ADONILDIEDAVGCONEJLÑSEDATNIAE_ON_APVRSDLE_ON_AIEU_EONI
MI_DXGSENTAPZEREQÑU_BMANEDTÑL_AVXDSQHF_EOUKN_FTMM_NEAAR_AVL
E_NITYJLEREONIAPVMINN_GPISUYNEAAA_EUABIAGSQ_QQQ_ETAZMREMMM
TE_IAPI__SEQUEEE_GP-----------------_EOILAAXZBRLMTET_ET_ET
LEIAARSLAREAR_ADQHÑTJ__ÑRUIURSTLIALCC_ETBBSZ_SQTCOEIEI_YEON
E_ONLI_ASHTEO_ASEJLTTMN_AGGVL_ONEAXUR_ASEDATN_EOTIAPN_REDTE
_EONUJSCO_GTRB------_ER_AÑLC_IDRSTAU_EO_U_EO_EO_AZLE_NIRSUL
_EOMIAUS_NTAZS_ZZI_AQEOLLE_AV_UREALE_RDPOEILTSMQ__BAPCLN_RD
MN_AUL_EOPDQEKHÑTDFL__BAHH_FBMN_F_BF_BR_YLAA_REO_HT_ES_EMRD
_EO-------------------_ES_ED_EO_SE__EO_R_EO_KLC_ESR
I_AMBSUYYLYBHLEBO_DVG_ERSQHTUE_ONLA_OENI_---------------
_EONI_APGVSEON_AVGOEI_S_EOTAPGO_ENIUCLE_DGPAI_EO
_EO_JLC_EO_IEON_ETLUCJLE_OIES---------MO_E_OEI_OE_R_EO_APG
_EO_E_CUSEOMSTUU_O_UUU_E_OGPVADQSHTE_CTR_ESSCYIMBAVPCE_OCLJ
O_E_ULCE_ON.
-E_O_CLYCUE_ONULE_O-VPON_EO_CLEO_YLC_EO
NEEOE_IE_ON_EOCCLJOECNYLCE_TDFFS_EFSMB_GCLE_--

By all appearances, therefore, the encoding algorithm is not based upon simple substitution (which one could have easily envisaged, considering the sequences of three equal letters which occur here and there).

The "Italian Ummites"

The Italian Ummites seem to be a very peculiar type of their kind, people who were representing themselves as Ummites, but who probably had very little in common with their Spanish counterparts (although, for instance, they maintain a strict formality with their names). These people were present in the Rome area for a very long time, establishing contact with a group of university students and some teachers, these latter now retired. They claimed to have built a huge base, 33 km under Veio (an Etruscan necropolis North of Rome). The following plan was drawn by one of these persons, and given to the boys they were in touch with:

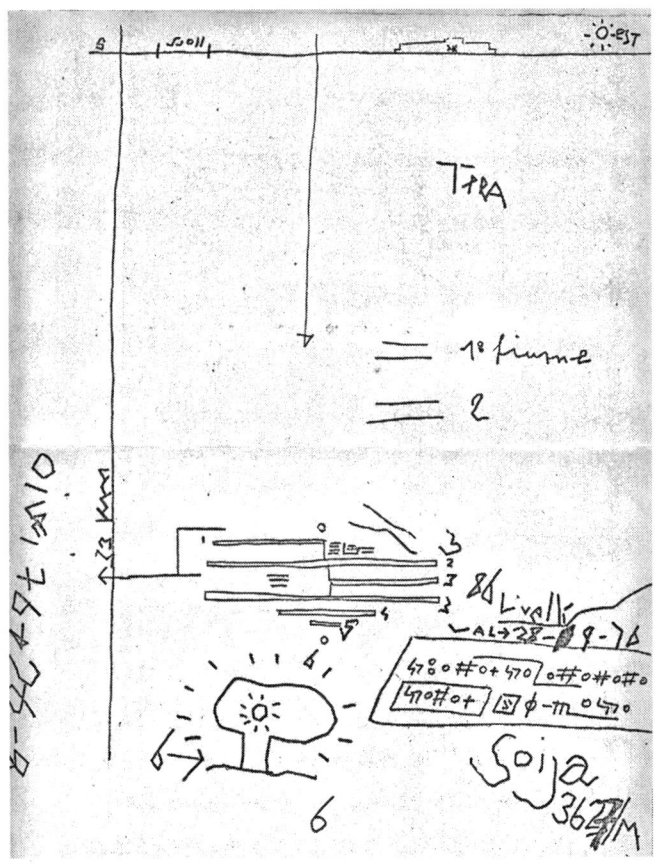

The formal writing style recalls, indeed, that of the "classic" Ummites, but some difference are noted, first of all the angularity of the letters, which is not seen in the classic texts. Moreover there are strange features: the base itself seems to consist of 86 levels, and two rivers run between it and the surface. It is not clear what the object in the lower right is, resembling a kind of mushroom with a sun inside it, and showing three digits looking like "6" all around.

In fact, years ago, a national newspaper featured a photograph, taken in Veio, showing the satanic number 666, sculptured on a rock.

Moreover, they claimed to have other, smaller, bases, just beneath the city of Rome, and even to own flats in its very centre. One of the boys actually recalls not only being taken inside one of these flats, but also to have been given its key.

Contacts went on for many years, also involving persons outside the group. Unlike the "Spanish" Ummites, they never sent one single letter. On the contrary their contacts were made by telephone, with very long conversations, most of which were tape-recorded by the boys, and then transcribed by typewriter. As I was acquainted with them, the (then) boys were so kind as to present me with some 7 cm (in width!) of sheets of paper, with the transcriptions. I believe I'm not going to commit any indiscretion by presenting here some excerpts, chosen totally at random (and in any case I am not going to present the full names of the people involved, but only their initials). Dates are in the Italian format: day/month/year.

> *4/1/1973*
> *It's 12:30 – Elko* (One of the "Ummites" – Author) *calls Fabrizio. Three whistles as usual; then:*
> *"We are going to meet tomorrow, myself, you and Gianni. Tell Gianni to choose a place he likes. I will arrive in a blue 500* (Author: A small FIAT car) *whose registration is Teramo* (In Italy, in those years, car plates contained the name of the district the owner resided in. In this

case the plate should have read TE, followed by some figures). *Bye, bye, bye."*
Gianni calls Fabrizio at 13:15, and they decide to meet at 16:30. Gianni will take his camera with him.
13:25 – Fabrizio calls Gianni to confirm.
14:00 – Elko calls Fabrizio:
"So, has Gianni decided?"
Fabrizio answers, then adds: "Elko, will I come on my car?" "No, you aren't coming, your double will come in your place. You'll go and have an ice-cream to the Zodiaco (A magnificent restaurant almost on the top of Monte Mario, a hill overlooking Rome – Author), *together with Antonello and Giorgio. But Antonello will have just a camomile-tea, because he isn't very well."*
14:10 – Antonello receives a phone call; only three whistles.
14:35 – Another call to Antonello: "OK".
14:45 – The telephone rings three time at Gianni's: it's OK.
Fabrizio calls Gianni to tell him he will not participate.
14:48 – Elko calls Gianni; the telephone call volume drops a couple of times, for "energy problems", and he apologizes.
The main topics are:
There is a woman of theirs in Rome, maybe we'll meet her;
Elko makes a call exploiting the energy of SIP (The telephone provider in those days), *and the sounds heard in advance are meant to verify whether the line is free, and it is possible to speak.*
He will come to have dinner at Gianni's: he only eats natural products; fish, vegetables, little meat.
He knows Gianni's parents, and is sure they are benevolent towards them.

4/1/1973
Other alien races are present in Italy. The Ummites know them; they are disgusted with the W56 (Another race,

which we are going to speak about at length about it in the third chapter), *who destroy their crafts; they only try to defend themselves.*
19:30 – Elko calls Gianni:
"*In Florence, Arezzo and Perugia there are the Elta V* (Yet another race - Author). *Their sun is called Xilias, and belongs to the Orion constellation. I shave my moustache when I'm travelling. I am short-haired, although I'm only 22. I saw people dancing, in an EUR* (A quarter in Rome) *resort, I've drank orange-juice, very good. I didn't dance, just watched. The article on ABC was awful* (A tabloid, in those years). *Mavi and Cear* (two of the Ummite people) *are to get back in January, or maybe February. Tomorrow you'll see the robot* (It's a biological copy of the body of a person) *of F.; tell A. and F. not to come, or otherwise you'll see nothing. We eat chicken. We know dogs quite well, and have them on our planets."*

5/1/1973
15:15 – Elko calls Gianni: "Get out", and hangs up. Gianni gets down to Medaglie d'oro street, near to the Esso gasoline station.
I go down. In a moment, from Tito Livio avenue, a dark blue 500 is descending. Its plate reads Teramo; I'm looking at it from the crossing; I see Fabrizio! It can't be him! The car turns to its left, and I go on taking pictures. Hoping that they come back, I go to the right; I hear their horn and I look. It's the two of them, and I take a picture. Elko is driving, he wears a tawny sweater, Fabrizio a white one, and dark sun glasses. I cross the street. They turn to look at me and slow down their car to allow me to take pictures. I show the letters I have for them, and they stop at the traffic lights. I give the letters to Fabrizio, who looks at me. It's identical! (only after this third time I've stopped suspecting a joke). Fabrizio looked serious. I took other pictures, also of Tito Livio Avenue and of the gasoline station. The robot

was not behaving normally, he was moving in fits and starts, like a puppet.

3/2/1973
"Fabrizio, are you ready?"
"Ready for what?"
"To see us, and, if you like, to take pictures ... Within seven minutes we will be over your house. Are you happy?"
"I'm quite happy, I'll get my camera: I have a colour film in it. I'm really happy that you give me this chance. That's a privilege ..."
"It's not a privilege, just logic to make you believe ... The place where you live is easier, because it's at a lower altitude. Get ready. Bye Brizio (Short for "Fabrizio")*."*
"Roger, I'll get ready. Call again."
15:07 – It's marvellous. It comes at high speed from the North, then it slows down, and flies over my house very slowly. I shoot many pictures, and they look as though they are willing to be protographed.
On the bottom of the disk there was their symbol. The craft had a golden colour, with a red dome above.

From these examples, chosen at random, it may be seen that "Italian Ummites" were indulging in manufacturing biological robots, clones of some of their contactees, and at times were organizing sightings: Fabrizio's house, in those days, was North of Rome.

They made different requests of the Roman boys. These had to undergo tests with unidentified medicines, plus more trivial tasks, such as growing a beard, or shaving it, attending specific courses at their University, making use of fictitious names (for instance the "Ibrn" that we are going to meet some lines later), and things like that.

Let me quote now the diary of an experiment:

23/2/1975
8:45 – I get to Veio, and in the usual exchange place I find a small package, with a label on it, stating that it must be open in the presence of A.; in this way the experiment starts.
11:00 – Applying the serum to our right fore-arms.
20:00 – Eight hours have elapsed. No apparent problems. At times it looks as if the corresponding place on the other arm was becoming irritated, but maybe it's just a psychological response. At times A. suffers a little pain in his left scapula. Lucid mind always. We are waiting for a confirmation of having applied the serum correctly. Breathing is normal. Body temperatures of A. and F. equal to 36.7 degrees (centigrades).
22:30 – "Hy, hy".
"Hy. We were wondering how the serum must be applied. We have just smeared it on our forearms."
"It's OK even that way. On Thursday you'll wash it away. Sodium hypo-chloride, and, if you have any, you may use also other reagents. Then, apply the powder, after having melted it with pure water ... distilled water will be OK. Especially over your veins. An exchange must be activated with your tissues, then with your blood, to modify your electro-magnetic halo."
"But what's the use of it?"
"To receive pulses. It will take effect within 18÷23 days. We make use of a substance (the eosin) that you already know, but do not use in the right way. We are going to transmit to you from 21 to 23 at night. It will be sufficient that you sit at the table, and write down, using the alphabet we have given you. You will feel pulses in your cerebellum, but remain relaxed. We have already done this experiment in Spain, but there it failed. If it works this time, no sintex (I do not know what this name means – Author) any

longer, because we'll be able to talk directly to each other. Bye for now."

Fabrizio, who was in a sense the charismatic leader of the group, had told me that "his" aliens were persons totally similar to us, except for their eyes, which were practically circular, without the angled end that characterizes ours. For that reason, when they had to walk around, they usually wore dark sun-glasses. They had a set of referenced points: the Zodiaco, at Monte Mario, from which a magnificent view could be admired, a self-service cafeteria in Cavour boulevard, Clodio square, and of course the necropolis in Veio.

Another transcription, about this last place:

21/8/1976
At Veio, myself, S. (A classmate of mine, during high school – Author), *and Ibrn, in the place named "La villa". Ibrn and I go down inside the vault, and soon after we see an almost unreal guy, wearing an overalls, who is lighted from behind him. He has to talk with Ibrn, so I go back. While climbing back, I see a robot, which I recognize as a clone of S.B.* (I hope that those letters do not refer to me! – Author). *Here and there S. and myself hear Ibrn's voice, and those typical of our friends. A few minutes have elapsed, when I hear S. shouting "It's them! It's them! I heard them! It's frightening!" He climbs to the surface, often stumbling, then runs to me, and while shouting "Let's get away, let's get away at once!" actually jumps into my arms, as a baby does when if it fears something, lifting his feet from the ground.*
Three days later Ibrn and myself go back to the same place, and find its entrance closed by tufa bricks.

As may be inferred, things were not always quiet. One morning (I was living in Pescara), at four o' clock in the morning, I was awakened by the doorbell ringing endlessly. Opening the door, I found two friends of mine, literally hysterical. A few hours before they had gone to Veio, and had apparently been targeted by some UFO's which had a good time terrifying them. So they had fled towards Pescara, seeking an help from me (what kind of help could I have given them?). I don't dare think of their mental state when they drove away along the Cassia (the road that joins Veio to Rome), and then along the highway to Pescara, exposing themselves and who knows how many other unconscious drivers to who knows what risks. Hopefully it was completely dark ...

Sometimes, looking after UFO's may be a daring experience!

Among the "Roman" Ummites, a girl, miss Swollha, stood out. Actually the "h" in her name was an invention of mine, because I believed I could hear a little aspiration before the "a", and she was rather perplexed by this. She was very active in Rome and outside, for many years, making telephone calls to a lot of people. What was spectacular about miss Swollha was her voice. It was a really warm and seductive one, to the point that several human wives became jealous of the woman who was calling their husbands!

My first "encounter" with miss Swollha took place one evening at my home in L'Aquila. I received a call from her, which began with her saying: "Hallo. I am Swollha, from the Ummites." I got rid of her in a few words, then I said to my wife (who was present): "The Martians are here again!" (please remember that I use the word "Martians" in a widest sense!). My wife replied that it was probably just a joke. The telephone rang again, and this time she answered, and heard Swollha's voice saying: "No, madam, it's no joke!"

I got many telephone calls from Swollha. One night, with Fabrizio and his girl-friend having dinner with us at my home, Swollha called, greeting the all of us, and expressing her opinion on the various courses of the meal! Another time I tried to persuade her to speak to me in her native tongue. However (as I was later able to understand only after distressing discussions) she answered that it was simply impossible, for her Italian speaking was the result of a computer that was translating her words into Italian, and for some strange reason it was not possible to override the machine – therefore her famous voice was a synthetic one!. A few minutes after the talk was over, she called again, this time actually speaking an unknown language. I was caught off guard, because I wasn't expecting that, and started to answer hawkwardly in Russian, because at first that was the language I thought I recognized. I enclose the text of the very short conversation (underlined vowels are the one with the tonal accent):

9/11/1977

Ah, è Swollha? (= "*Ah, it's you, Swollha?*")
Nja idljepor ifakude nje riski sikurajote o shte ketch.
Si. (= "*Yes.*")
Ajo o shtetarnian bah vertetashto to asht a prjet shti nishti.
Я полагаю что говоришь по русски. (= "*I believe you're speaking Russian*")
O mashkaro set ta vjonde mesh k optaim muan gadamt, ekeni tengadjamt e k liepr.
Да. (= "*Yes*" - ???)
Panim shto a ti o ljepor.
Но я не понимаю ничего. (= "*I don't understand anything*")
Tufari kadu pret to li kurajo tabut ljepori mun go fridjem o li koshem pertjemt.
Но это не русски. (= "*This is not Russian*")
Ti, ti accontenteremo, (= "*We are going to satisfy you*")
D'accordo. (= "*Roger*")
Se avrai discrezione. Molta. (= "*If you'll be very discrete*")
Che cosa vuol dire? (= "*What do you mean?*")
Ti accontenteremo se avrai discrezione. (= "*We are going to satisfy you if you'll be very discrete*")
D'accordo. (= "*OK*")
Miznat.
Понимаю. (= "*I understand*")

Actually I had cut a poor figure, with my strange answers. A few days later, one of the boys in Fabrizio's group received a telephone call, and a long conversation took place as usual, during which the Ummite said "Stefanino (it's my nickname) doesn't speak Russian well"!

To conclude, one of the boys (who was studying cartoon animation at prof. Di Girolamo's office) has presented me with this pleasant cartoon, referring to the Etruscan necropolis; the words spoken by the alien sitting on the disc's rim read: "Is everyone dead here in Veio?"

Stellar Problems

According to the Ummites (the "Italian" ones, in particular), the following stars host civilizations, besides, of course, Wolf 424, from which they claim to be coming. Actually, aside from the super-giant star Alnilam, at an enormous distance, and the two Wolf stars, all the others are considered by conventional astronomy as possible candidates for planetary systems. The first one, Wolf 454, is probably a mistake on the Ummites' part, as I've not been able to find a star so named in our stellar catalogues.

Name	Name	Spectral Class	Absolute brightness	Distance (l. y.)	General characteristics
Wolf 454		?	?	?	?
Wolf 424		M7	14.2	14	binary
ε Orionis	Alnilam	B0	-6.39	1,340	Super-giant
61 Cygni		K5	7.07	11.4	binary
τ Ceti		G8	5.69	11.9	yellow dwarf
E Ursae Minoris		G5	4.2	347	triple
δ Pavonis		G8	4.62	19.9	yellow dwarf
70 Ophiuchi		K1	5.8	16.6	double

The claims of the Ummites regarding their home star received an apparently serious blow in the early 90's, when the Hubble telescope resolved Wolf 424 into two stars, therefore into a binary system:

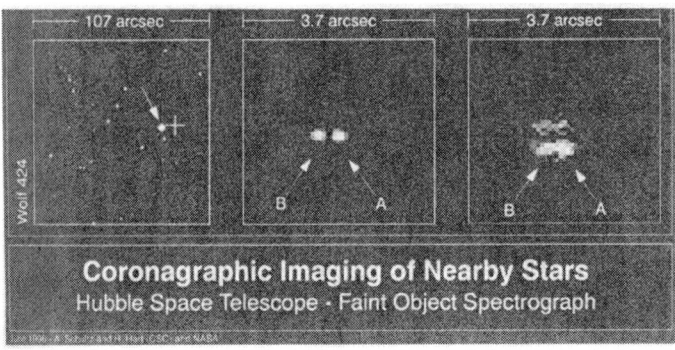

Coronagraphic Imaging of Nearby Stars
Hubble Space Telescope - Faint Object Spectrograph

These are the characteristics of the two stars:

Reciprocal distance	3.1 astronomical units
Period of revolution	16.2 years
Masses	0.14 e 0.13 solar
Diametres	0.17 e 0.14 solar

We are dealing here with a couple of red dwarfs, whose reciprocal distance is, roughly, as much as the asteroids belt, between Mars and Jupiter, is distant from the Sun. Actually, if there were a star in that position in our system, Earth would have an erratic, wandering orbit. But we must note that the masses of the two stars are a little over one-tenth of that of our sun, and so gravitational effects are similarly reduced.

It has been claimed that the Ummites have never said that their star is a binary one. Indeed it is not so. In Ribera's book "El misterio de Ummo" (1979) they state that their system is a trinary one!

> *The possibility cannot be eliminated that the star you call WOLF 424 is one of the OOYIA (small bodies) that we have codified as:*

꓿꓿꓿

꓿꓿꓿

The distance from our Sun of these two stars should be, according to them, 20.7 light years and 0.62 light years.

But, put this way, things make no sense. The distance of the first star from Iumma should be greater than our sun (about 14 light years, as we have seen). Even if we think that they have made a mistake in placing the decimal point (that is, for instance, 2.07 instead of 20.7) it would not be convincing, because the distances would be any way too great. Probably the mistake consists in their having used the term "light year". It must be a different unit of measurement.

The Hubble telescope has found the distance between the two stars as 3.1 astronomical units. If we pose:

0.62 "light years" = 3.1 astronomical units,

then:

20.7 "light years" = 103.5 astronomical units,

and that would make perfect sense.

Regarding the second star, if it were some tens of astronomic units away from Iumma, orbiting in a plane rather perpendicular to the plane of Ummo's orbit, not only would the wandering problems be drastically reduced, but this would explain a report (otherwise incomprehensible) about the origin of their cosmology (which I do not quote here, because it's too technical).

Ummite Geometry

While their arithmetic is similar to ours (although they use the base 12 instead of 10), their analytic geometry is quite different. We use an orthogonal Cartesian reference system, and identify each point on a plane via two numbers, the measures of its relative distances from the two axes.

Ummite geometry is based upon two poles, A and B. Any other point C on the plane is identified by the values of the angles α and β present in the following scheme (clockwise for the first one, anti-clockwise for the other):

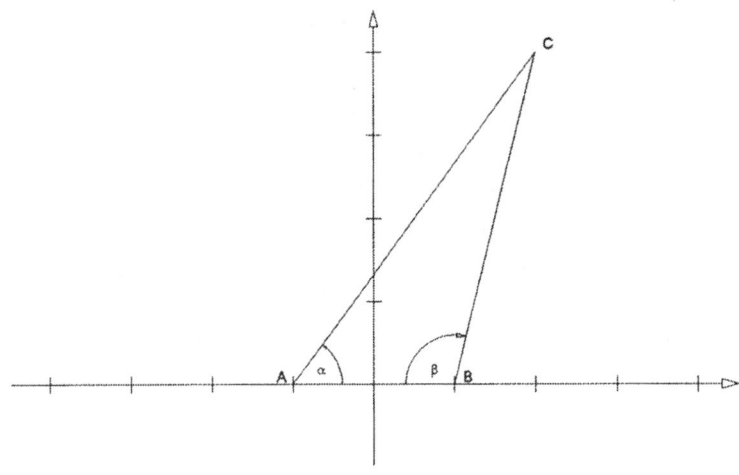

Let w be half the distance between A and B, it is not difficult to find the equations that correlate the two metrics:

$$x = w \cdot \frac{\sin\alpha \cdot \cos\beta - \cos\alpha \cdot \sin\beta}{\sin\alpha \cdot \cos\beta + \cos\alpha \cdot \sin\beta}$$

$$y = 2 \cdot w \cdot \frac{\sin\alpha \cdot \sin\beta}{\sin\alpha \cdot \cos\beta + \cos\alpha \cdot \sin\beta}$$

Both these formulas may be reduced to:

$$x = w \cdot \frac{\sin(\alpha - \beta)}{\sin(\alpha + \beta)}$$

$$y = w \cdot \frac{2 \cdot \sin\alpha \cdot \sin\beta}{\sin(\alpha + \beta)}$$

The inverse equations are evidently:

$$\mathrm{Tg}\,\alpha = \frac{y}{x + w}$$

$$\mathrm{Tg}\,\beta = -\frac{y}{x - w}$$

(one has to take care of the periodicity of the tangent).

It's interesting that a linear equation in α and β represents a circle:

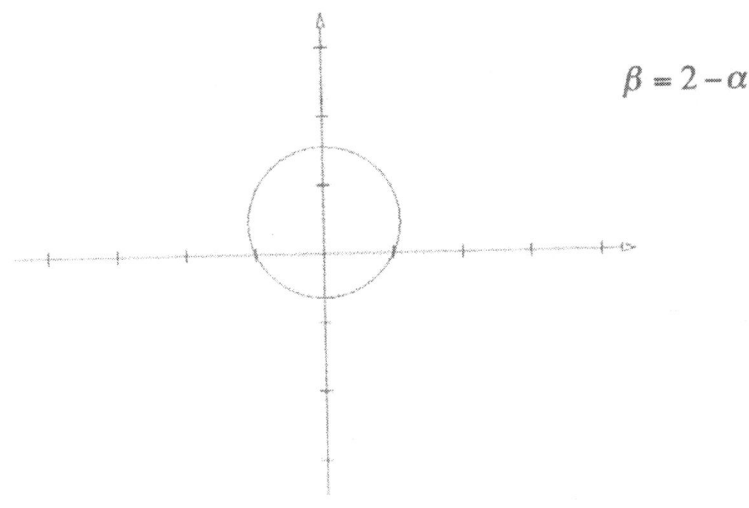

$\beta = 2 - \alpha$

It may be easily demonstrated. If we write the equation this way:

$$\alpha + \beta = \gamma,$$

if γ is constant, that means that in a generic triangle ABC the angle in C is constant, therefore C lies on a circumference that goes through the points A and B.

Its radius is given by:

$$R = \frac{w}{\sin\gamma},$$

and the ordinate of its centre is:

$$h = -w \cdot \frac{\cos\gamma}{\sin\gamma}.$$

It is first necessary to compute its radius: let's consider that the angle γ remains constant also when the point C coincides with A; in this case, γ measures the angle between the tangent in A and the half-line $A \rightarrow B$, and is therefore equal to the angle between the half-line $O \rightarrow D$ and the half-line $O \rightarrow A$, O being the centre of the circle, and D the central point of the segment AB:

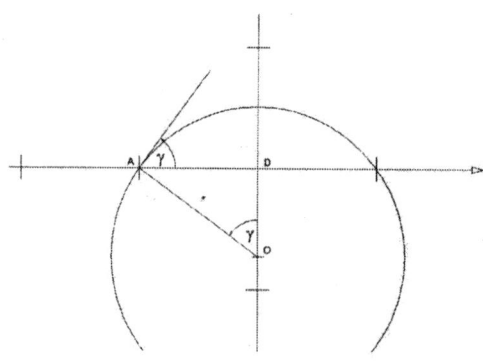

For instance:

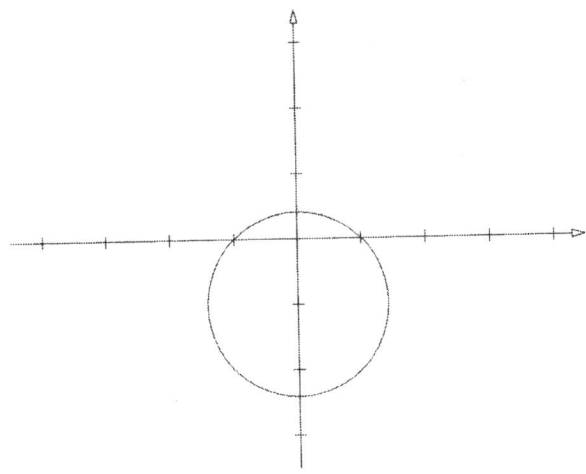

The equation of this circumference is::

$$\alpha + \beta = \frac{\pi}{4},$$

and therefore:

$$h = -w$$

$$r = w \cdot \sqrt{2}$$

It is not difficult to demonstrate that the equations of the kind:

$$\alpha - \beta = \gamma,$$

with γ constant, represent hyperbolae, whose asymptotes are oriented in various ways.

It is amusing to see what happens when γ = 0. The resulting locus is given by the usual pair of the orthogonal coordinate axes! Indeed, if α

≠ 0, the resulting function is the ordinate axis, because the points *C* are the top vertex of isosceles triangles, therefore they lie over the straight line *x* = 0.

If α = 0, it follows that β = 0, and the triangle *ABC* degenerates into the straight line *y* = 0. Really amusing, even if it is easy to be imagined. Indeed, in our Cartesian metrics, the reference axes are also the limit of a family of hyperbolae.

It would not be easy at all to extend this criterion to three dimensions, were it not for a seemingly queer concept that the Ummites introduce while describing space. According to them, the space is composed by a threefold infinity of points (thinking in three dimensions), but to each point a threefold infinity of angles is associated, as is partially exemplified by the drawing:

We have already seen that the locus of the points that see the two points from a constant angle is a family of circumferences. Extrapolating into three dimensions, we have a family of toroids. If we impose a function to the variation of these angles (that they call IBOZOO UU), any kind of a structure may be originated. They explain:

> *We are aware that the WAAM (cosmos) is constituted of the network of IBOZOO UU. We imagine the space as*

a set of combined angular factors. Were you UUGEEYIE (CHILDREN) attending a school, may be we could make use of a silly example: the Universe is like a swarm of dragon-flies, whose wings are at different angles. But such flying insects have a Mass and a Volume (at least in our mind). The IBOZOO UU is not a corporeal particle with a Mass. As a first conceptual approximation we might think of it as a network of oriented Axes: the important thing with this network are not the axes themselves, but the angles they form.

In other words, what our senses interpret as a linear entity, for instance a straight line, or, as you say, a Linear Scalar, it is nothing else than an OXOOIAEE (chain of IBOZOO UU). We receive a deceptive image that is the result of the work of our brain in integrating a synthesis of this Network of IBOZOO UU (that in the WAAM are indeed to be found in disorder, and without a specific location).

Who knows why, but at this moment they change abruptly to a subject that appears to have nothing in common with geometry:

But at this moment you are in an ascending period, and we in our descending one, that may be represented by the following diagram:

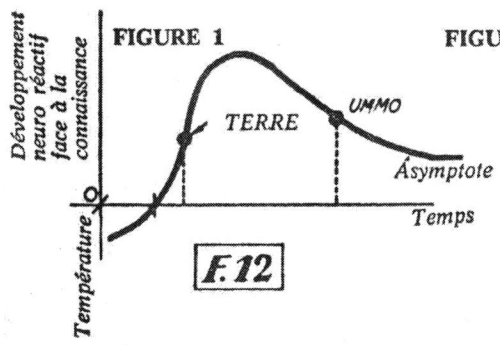

On Ummo we have created a science, or theory, called AYUBAAEWAA. Mathematicians will be able to

understand us if we say that it is equivalent to what you call the Set Theory. The difference is that the AYUBAAEWAA is oriented to examining Sets with Mutual Relationships among themselves (what you call Network Theory and Set Theory are just simple sections of our generalized theory).

About the speed of light in a vacuum, that's what they maintain (some passages are from "El Pluricosmos"):

You think that the maximum speed a sub-particle may attain in the WAAM (Cosmos) is 299,780 km/sec (the speed of light), and you consider this speed as "CONSTANT" ... It's the same speed we have measured ... in this three-dimensional environment. But it is sufficient to change the environment, or three-dimensional reference system, to see this LIMIT SPEED to change greatly, up to a point where the only reference that might reflect the axes changing is just the measure of this constant speed C.

This way we have a sequence of values:

$$c_0, c_1, c_2, ... c_i, ... c_n$$

that extends itself from $c_0 = 0$ to $c_n = \infty$, each one of which represents a well defined reference system. What physical medium could allow us to understand which reference point we are involved with? One is sufficient: When there is no mass perturbation, to measure a time interval ... This interval is the one required by the movement of a quantum of energy along a standard distance.

At the moment you succeed in controlling, as we have achieved, the homogeneous inversion of all the sub-particles of the human body, or of any other object, you will assist to the passage from a three-dimensional SPACE to another one, still three-dimensional, but different from the former.

The concept of "inversion" of the axes of the IBOZOO UU is, to them, the base mechanism that allows the change from one of the c_i to another. However they do not give us many details on this subject.

Appendix

My friend Riccardo Di Prinzio, founding member of the cultural association UFOObserver in Pescara, and member of the amateur astronomers group in Chieti, has given me his interpretation of the Ummite symbol. Up to now, it is the first (and only) attempt to find a meaning for that strange design. However, his explanation is very concrete, and based upon astronomy. With his permission, I have decided to include his theory in this text, and let him speak.

The Ummite symbol

I've always been curious of "Ummites", whose symbol, in particular, fascinates me, and I invite you to reflect on it. This symbol (which appears on the bottom of some flying saucers) is a strange combination of a cross and a letter "H". But what does it mean? Does it represent anything? It may be a mere coincidence, but in astronomy this symbol is almost identical to the lines of shadow drawn by a gnomon (a pole inserted vertically into the ground, which in ancient times was used to reckon the time thanks to its shadow, a kind of sun-dial) that, during the year, marks important astronomical events.

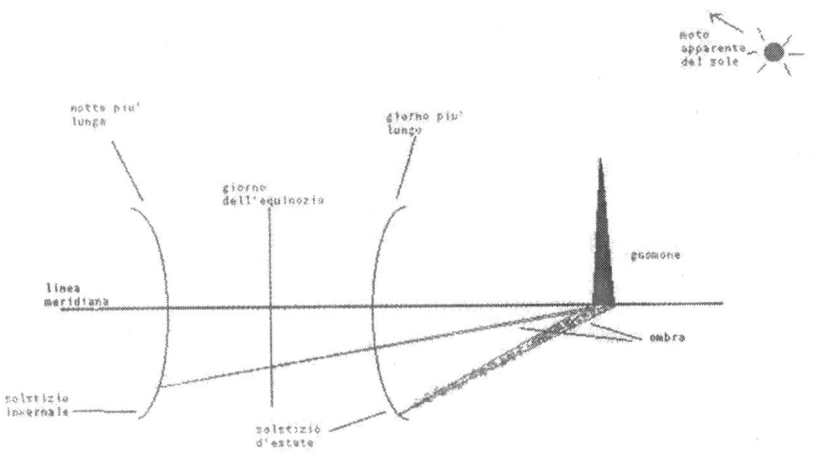

On a sunny day, if the line described by the shadow of the top of the gnomon forms a curve whose concavity faces the gnomon itself, that means that the day is longer than the night, and vice versa in the opposite case.

Obviously, going from the period when day is longer (spring – summer) to the period when night is longer (fall – winter), there will be a moment when day and night will have the same duration. This happens on the equinox, when the line drawn by the top of the gnomon with its shadow is a straight, perpendicular to the meridian line, the one along which the shadow is the shortest one, of all the days of the year.

The shadow line furthest from the gnomon will be that drawn during the winter solstice, and the nearest one during the summer solstice. On the spring and fall equinoxes, as I have already said, the shadow lines will be perpendicular to the sun-dial, and they will mark the curvature inversion of the lines.

The meridian line, then, divides the day into two parts, marking mid-day, which is the moment when the sun is at its highest.

Using only these three lines, we are able to sub-divide the year into its four seasons. Isn't it fascinating?

But do the Ummites have anything to do with all that, or is it just a coincidence?

Conclusions

I hope it has been evident that the story of the Ummites is very varied, full of information, much of which reveals a very high level of technical and scientific contents. I have presented here just a few examples. What may we say?

UFO scholars usually tend to see the Ummo affair as a gigantic hoax, but without offering any proof of that (the most severe critics referring to the clouds of San José de Valderas!). Of course, there is no actual proof of the validity of the story, beyond what the interested characters themselves have related. On a couple of occasions, they even said to have artfully interspersed here and there false information, easily debunkable, when they realized that human interest toward them was growing beyond what they could accept.

Surely, if it was all of a prank, someone would have had put together a team of people expert in the most varied branches of knowledge, and instructed them to envisage a scenario beyond our resent-day culture, but still plausible. Such scientists would have therefore been the originators of the many reports. A world-wide network of collaborators would then have taken care of mailing them from the most remote places to persons carefully selected, such as the time I received greetings from Leningrad, showing that they had done research into the name of my father! Obviously such an operation should have cost a huge amount of money (at least in stamps). *Cui prodest?*

The ones who maintain it was a fraud implicate the secret services of the main powers on Earth, the CIA and the KGB, but they also suspect mysterious occult governments (which are much more improbable than the gentlemen from Ummo), to covert entities (idem), or even to Jesuits (?!?). Setting aside such ravings, one can't understand why the CIA or the KGB should have played, at the height of cold war, such a colossal trick, which continued even after the collapse of the Soviet Union, and of the KGB itself.

A minimal amount of common sense should make us believe that the story is true. Unfortunately, this hypothesis also presents several difficulties.

By all appearances, their first landing (Cheval Blanc) was planned in a thoughtless way. If the scouts were really forced to face a hostile reaction from the part of Earthlings, would they have been able to defend themselves, even with their superior technology? Let us remember that the only data they possessed had been taken over Switzerland, of a few minutes duration. They could not understand any of our languages, so listening to wireless for such a short period had been of no use. The adventure with the herdsman (herdsboy, indeed!), the theft in the farmhouse, everything shows improvisation and dilettantism. Also the story of the "severed hand", beyond its sensationalism, remains rather mysterious - for what reason had the two alleged Norwegian physicians leased vaults (paying very high rent for them), with the obvious risk of involving the owner, with the domestic staff that was constantly coming and going? As they looked to have no problems with money, why didn't they simply buy some structure off the beaten track, where they could devote themselves to their research free of interference?

The economical aspect is another mystery. We may suppose that, once having discovered the role of money in human society, and the higher intrinsic value of some goods (gold, precious stones, heavy metals, may be even drugs), they started accumulating money by selling shrewdly here and there some of these things that are so valuable to us. But to maintain an expedition composed on completion of hundreds of people, they should have taken care to spread their sales pattern over a wide area, in order not to create any suspicions among the police of the various nations involved in the process, or in the buyers themselves, and this process should have required a plan much more careful than the one they had exhibited. It's funny to think of a group of Ummites, acting as drug pushers in the streets, but it is doubtful it happened that way, because they would surely have been swept away by the local Mafia.

We might also think of a first, huge, investment, followed then by an intelligent administration of the capital thus obtained. On our planet, the richer one is, the easier he increases his wealth. Ummites could have sold, at a high price, some innovative principle to one, or more, big companies, and could then have made further profits on the money obtained this way. In principle it could have been possible, and actually those years did see some inventions totally unaccounted for. But should we believe that our friends were able to act as financial gamblers, a talent for which where they were obviously totally unprepared, and which their need for total secrecy would hace complicated immensely?

Or we might think of the various small dictators (Africa, South-East Asia, South America), who typically have their people starving, while not hesitating to spend a lot of money for weapons, in environments obviously beyond any legal scrutiny. But could the Ummites have been able to produce weapons on a large scale, without being noticed (and, again, without generating any reaction by the Mafias involved)?

Then, aside from these problems, it's difficult to understand the strangeness of their communications. If they wanted to remain covert, why did they send tons of paper around the world? And, even if we decide to accept their desire to make contacts with some precisely selected persons, why they were one-way only (with the exception of "Italian Ummites")?

Probably the story is neither white, nor black, and the real truth must remain somewhere in between. Any way we haven't the least idea where to look for it.

¿Quien sabe?

Notes

1. The only analytical equation I know of a square is: $\sin x = \cos y$
2. Just 15 years after the transmission: their ships must have moved at 1.5 times the speed of light!
3. Up to now, nothing has been found (at least nothing has been told on this subject!)

Bibliography

M. Castello, P. Chambon, I. Blanc: "La conspiration des étoiles" – R. Laffont, Paris,. 1991;

Renaud Marhic: "L'Affaire Ummo" – Editions Heimdal, Aix-en-Provence, 1993;

Juan Dominguez Montes: "El Pluricosmos" – Librería Agora, Malaga, 1983;

Jean-Pierre Petit: "Enquête sur les Ovni – Voyage aux frontières de la science" – Albin Michel, Paris, 1990;

Jean-Pierre Petit: "Enquête sur des extra-terrestres qui sont déjà parmi nous" – Albin Michel, Paris, 1991;

Antonio Ribera, Rafael Farriols: "Prova dell'esistenza dei dischi volanti" – De Vecchi, Milano, 1972;

Antonio Ribera: "Gli UFO – processo con testimoni" – De Vecchi, Milano, 1975;

Antonio Ribera: "Chi ci osserva dagli UFO?" – De Vecchi, Milano, 1976;

Antonio Ribera: "El misterio de Ummo" – Plaza y Janes, Barcelona, 1979;

Antonio Ribera: "Le véritable langage Ummo" – Editions du Rocher, 1984;

Antonio Ribera: "UMMO: La increíble verdad" – Plaza y Janes, Barcelona, 1985;

Antonio Ribera, Wendelle Stevens: "UFO Contact from Planet Ummo" – Privately published by Wendelle Stevens, Tucson, 1985;

Antonio Ribera: "Ummo informa a la Tierra" – Plaza y Janes, Barcelona, 1987;

Antonio Ribera: "El envés de la trama" – Plaza y Janes, Barcelona, 1987.

The History of Amicizia

First foreword

I met Bruno Sammaciccia in 1963, when, together with the journalist Bruno, I travelled towards Pescara to meet him at home. It was winter, and I remember that our train, in Sulmona, was travelling slowly between two walls of icy snow.

We arrived in Pescara at night, and Bruno Sammaciccia welcomed us with a sincere warmth. From that moment, without being aware of it, I entered the "W56 saga", although as a peripheral guest.

At a first sight, Bruno Sammaciccia looked like anyone else. But, getting in touch with him, even if only occasionally, one couldn't but realize how much this man acted as an important and necessary catalyst in situations where such different realities were mixing together – our Earth entity, and the alien one, that of the W56's and that of their opponents, the CTR's.

This chapter, which I have read and pondered, and which in some way involves us in an uncommon experience, is almost a manual about the synergy between earthlings and alien entities, beings who together discover the value of a peaceful way of living together, and who cope with activities that require a reciprocal faithfulness in "thoughts, works and endeavour" (Author: it's an old Italian maxim), activities meant to create here, on our planet, a more active social environment, based upon what was called "Amicizia" (Author: Italian for "Friendship"). It lasted several years, from the first contacts in 1956.

Bruno Sammaciccia, having being enabled to communicate with the W56's, shows us in an effective way some of the components of the alien group, or, better put, of the groups which in those times were operating in our country. Many bases, under the Appennines (Author: the central Italian mountain chain), or under the lakes in Lombardy

(Author: a region in Northern Italy), are described with simplicity, as if their existence were logically acceptable.

Some of Bruno's best friends had the privilege of entering these bases, and becoming aware of the reality of alien operating centres in Italy, and of meeting persons whom the tale depicts with vividness. He describes their interaction with the every-day lifestyle of earthlings, whose habits at times they liked to practice – eating, drinking, even smoking! He comments on the range of their physical heights, going from one meter to three and a half and their interest to the history and present of a humanity they were observing with interest and respect, although they did not fully agree with its actions and intemperance.

This book, based on memories of experiences covering a period of many years, is charming above all concerning the contact, first, and then the coexistence of humans and aliens, working toward a single goal. Dimpietro, Sigir, Sajù, Itaho are names that I have often heard. Others, whose names appear in the book, are Giancarlo, Gaspare, Giulio, and so on. I have met some of them in person. The tale goes on at varying speeds throughout events which may appear incredible, but repeated reading of what Sammaciccia describes with simplicity and in great detail, makes one feel as though it is a re-enactment of the events, with the Author's closest friends, just like a modern Anabasis, written by a modern-day Xenophon. At the end of the book, the reader faces a reality in which two different groups, are violently confronting each other, and fighting for control of Earth with actions and reactions reminiscent of "Star Wars". On one side, a confederation of free planets, on the other side a technological dictatorship aiming to dominate our planet, beginning with the control of the minds and the wills of its inhabitants. One wonders whether Lucas was perhaps thinking of similar events while making his movies.

Bruno Sammaciccia is no longer with us. His mind was lucid up to his last day. Giancarlo, his closest friend, had died before him. In November, 1978, the CTR's were able to enter and destroy most of the W56's bases, including the largest one, which stretched from Ortona to Rimini (Author: two towns on the coast of the Adriatic sea, some

235 km apart from each other), and from the center of the Adriatic to the centre of Italy. I leave to the reader to imagine the catastrophe that hit the W56's; no cruel description could equal what they themselves had forecast: "You'll see waters rising. They will be boiling over the place where we have built our big base." It is sufficient to read newspapers accounts of the time about sea waves going berserk in the central Adriatic, about the unexplained luminous phenomena, and about the terror experienced by fishermen trying to flee the area in their boats. The coast-guard radars recorded strange echoes. In the meantime, a spectacular UFO activity took place.

The national press published alarming statements, but no rational explanation could be found for the situation, which lasted a couple of months (Author: towards the end of this book, a description this strange phenomenon).

I stop here, moved and thoughtful. I too was in touch, at a minimal level, with this reality, and tried to get rid of it, without siding with one faction or the other. I do not deny my experiences, but I believe that the future of this planet depends upon its own native inhabitants – recognizing, of course, the need to emerge from our ignorance about the laws that govern the universe.

Paolo Di Girolamo
Roma, June the 13th, 2005

Second foreword

This chapter contains but a small fraction of a very strange UFO phenomenon, an ongoing story in central Europe for some forty years, of which very few people are aware, at least up to now. In Italy, the persons most deeply involved in this experience have named it "Amicizia". In Germany it was known as "Freundschaft", in the former Soviet Union as "Дружба", and in France as "Amitié". It is truly astonishing that so many people, mostly unknown to each other, chose the very same word to refer to their experience.

One of the chief characters within Amicizia was Bruno Sammaciccia, a man very active in those times. A lot of other people, literally hundreds, entered this story. As it started in 1956, it is not surprising that over the years many of them passed away. Such was also the case with Hans, a German friend of mine who died not long ago; I do not quote his family name, because he had asked me to conceal it.

This story is very unusual in many ways. First, it covered a span of time of more than 40 years. Second, people from at least five different countries were involved: Bruno speaks about Italy, Switzerland, Austria, and I might add Germany and France. Third, all of us were moved by the deep morality and sincere humanity on the part of the aliens. These were people who simply could not imagine doing any evil to anyone, people who were liking to eating well, to drinking, even to smoking, who were enjoying playing violins and tennis, driving luxurious cars and executive airplanes (in the 70's, when very few people in Italy could own a personal plane), who were in deep love with us and with our planet, not as superiors but as true friends, or older brothers. They lived most of their times in their huge underground bases, but some of them lived among us, inside our society, playing every kind of roles in it. One was an university researcher, another one the managing director of a rather important textile company in the center of Italy, a third one was a senior manager in one of the largest German Telecommunications (TLC) companies, and so on. But the most important difference in all the contact stories I have read about consists of the role we earthlings played in the eyes of our counter-parts from so far away. They stated several times that we, the poor dwellers of this distressed planet, enjoy almost unique psychical abilities, while being totally unaware of that. They said that this peculiarity was one of the main reasons for their coming here. Often it was even funny for us to watch these people, with such fantastic technology at their side, begging our help in particular operations against their enemies. And, when the story came to an abrupt ending, it was because of a fault on our part, when most of the earthlings became frightened, or were simply deceived by the enemy, and ceased to support our friends. This development was enough to endanger them, reducing their superiority against their enemies, so that they had to stop their operations here and retreat back home, promising, however, to return in the future.

It was an explicit decision by Amicizia people to keep everything concealed under the strictest secrecy, and there were very good reasons for that. Actually, once in a while something would have emerged publically, but always in a vague and uncertain way. Many European UFO scholars were aware that something was happening, and they felt that it might have been something important. But nobody, outside our group, has ever had even the slightest idea of how big and how important it all was. Now, after so many years have elapsed, now that only a few persons survive, Bruno has decided to make his experiences known to scholars, and so he asked me to be his ghost-writer; I was to receive his memoires, and add them to a book I was already preparing. So, aside from what I had already written by myself, my role now involved spending many afternoons at Bruno's, recording his memories, then, in my studio, listening to the tape, transcribing the data on my computer, trying to organize the concepts into some logical sequence. At the same time, from Germany, Hans, having heard of Bruno's decision, decided to do the same, and sent me a file with his experiences, so that I could merge it with the other tales.

I must say that I did not welcome the request by my two friends, because I believe that such things should remain concealed. It's not easy at all to tell such astonishing stories, facing the obvious risk of being taken as mythomaniacs, crazy people, hiding behind one's academic credentials and successes at work. Nor is it easy to present mankind with such a tale, who does not seem prepared to accept it, or even to show interest in such things. The big project our friends had suggested to Bruno, and to which he has devoted all his resources, was actually intended to overcome this mental gap. But unfortunately, as you will see in the following pages, in the end the thick-skulled earthlings got the upper hand.

As you will see, my task was not easy, because my friend did not make an ordered sequence of reports, but was often "leapfrogging" from one concept to another, in the way we Latins call "Pindarics". At times I have arranged his memories chronologically, while at others I preferred to leave them as they were told to me, to grant them their spontaneity. Unfortunately, my English is too poor to express the liveliness of some Italian expressions my friend often uses, but any way I believe I have tried my best to capture their flavour.

Writing this chapter was a most interesting and pleasant experience for me, both because I personally learned things of which I was totally unaware, and because I took pleasure in remembering feelings that I had shared with Bruno and the other actors within the group we had named Amicizia.

Unfortunately, just after the manuscript was completed, both Bruno first, and Hans a few months later, passed away. This posthumous collection of Bruno's memories is intended as an homage to his mental stature and to his profound sense of humanity.

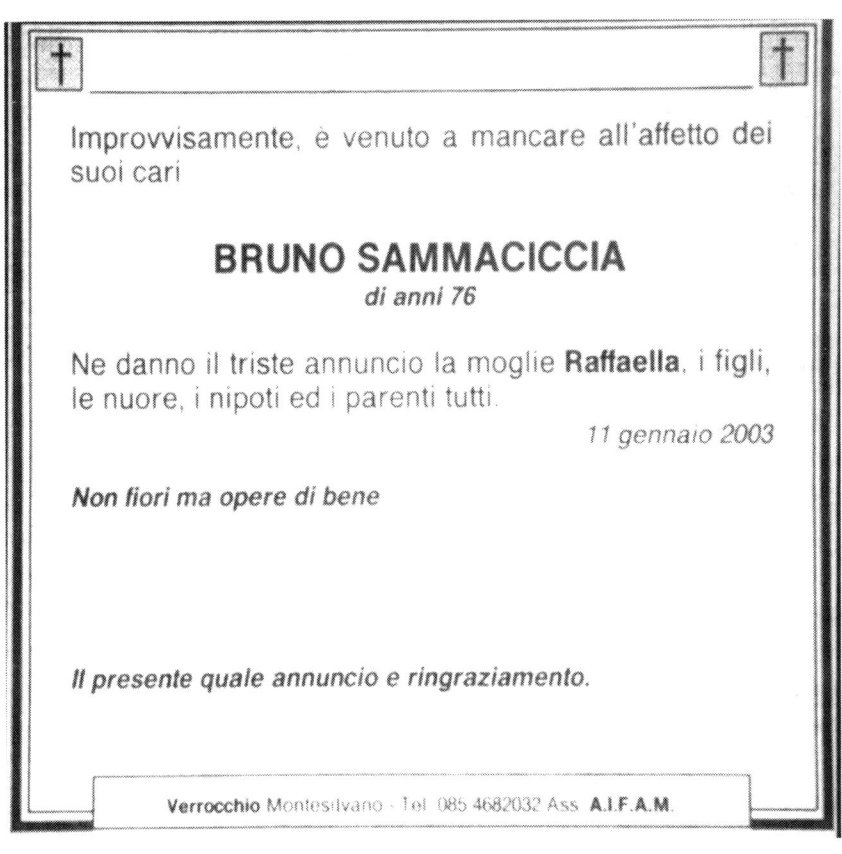

I hope to be pardoned for having deleted from this obituary all the references to Bruno's family. I do not wish them to suffer any annoyance from this book. Also my German friend deserves my full respect, for he was a colleague of mine, a friend, and above all a very good man.

One of the question Bruno and I posed to each other, while thinking about writing this chapter, was what to do with the names of the persons involved. Although most of them are no longer with us, it would not be fair to present them using their actual names. On the other hand, many of them played important social roles in everyday life. Therefore presenting them under their true names would offer the reader convincing evidence that what is being told in this book is not foolish at all, but is in fact the story of what has actually happened to a group of people among which was represented every possible social level and cultural point of view. Bruno received four international awards *honoris causa*. I myself am very proud of my personal achievements. Giancarlo, another major figure, was just an accountant. I do not mean to disparage accountants, for my father was also one, but I want only want to stress how wide was the spectrum of professional and cultural backgrounds among the people of Amicizia (even one, or possibly two, Nobel laureates were involved). In order to protect the privacy of the persons quoted here, I had initially proposed Bruno to write a book devoted specifically to Amicizia, directly in English, with an American audience in mind, where identifications could probably not be made out so easily. Therefore this chapter was at first written in English, then re-translated into Italian, (now back to English again!) as part of the first edition of my book, and the people referred to are presented using just their first names.

As I have already mentioned, Bruno was a man with a humanistic background. He understood nothing of technology, nor was he interested in it. I am totally at the opposite, having a strong scientific background, over which I have evolved a humanistic skin. I have met many persons who, though having a scientific background, have opened their minds to humanistic culture, but, as far as I can remember, nobody among the people I know has ever gone the other way.

This is a real problem with Bruno. He is going to describe, through my computer, a shocking experience, in which technology was most important, but he did not recognize that. For instance, he tells of objects being "beamed" away via remote control, that is, being disintegrated at the place they were, and being reconstituted someplace else. To him it was just normal that something like that could actually happen. Therefore, Bruno will seem to tell a kind of a fairy tale, where transcendental things happen without evoking anything but a slight surprise on his part. Bruno's real interests were in the moral meaning behind the experiences, in the feeling of Amicizia (friendship) which pervaded everything, in the spiritual nobility shown at all times by our friends, and, just to be a bit malicious, in finding confirmations of his attitudes when considering our reality, and then to find a surprising way of overcoming that point of view.

I am certain that this book will come as a real surprise to most scholars, because for a long time in Italy and in Germany some UFOlogists occasionally tried to get in touch with Amicizia people, often at no avail. Bruno was often sought out, but he almost always refused any contact. On just a couple of occasions he agreed to have a talk with someone, which he regretted having accepted, because afterwards rumors arose that he was a swindler! He was even accused of, and was tried on, charges of abuse of trust (he had been denounced by two persons who blamed him for their being disinherited by a wealthy relative), of which he was cleared. Preposterous! Among Bruno's close friends were a general of the Carabinieri (the traditional Italian Police army), as well as bishops, prelates, managers and businessmen, and, by the way, myself.

Of course this history is totally unbelievable, nor does Bruno want to offer any evidence of that, except for a few pictures; I am aware that pictures may be fakes (as you will see, one IS a fake!), so that they are of low reliability as evidence. I can only underline the fact that most of them are Polaroid shots, and that in those years it was not an easy task to counterfeit Polaroid pictures. Aside from that, there were many attempts made to apply some of the physical principles suggested by our friends to our technology. But, as far as I know, there was only one success. It must be remembered, indeed, that Amicizia (the group of earthlings) was not a homogeneous set of people, but it consisted in many people, each with his own culture, needs, expectations and reactions, that some leaders (Bruno among them) tried to organize into a unified group. It was an attempt that failed badly.

Bruno had no need to try to convince anybody, nor was he interested in doing that. Despite many rumors about him, my friend never attempted to gain profit from this story. On the contrary, from the financial point of view, he spent a huge amount of his own; he even had to build a very large villa, following the designs made by our star friends, receiving there many guests there at his own expenses. Some persons were his guests for years, free of any charge. When the story ended tragically, Bruno lost billions of Liras (millions of dollars) literally overnight, and suffered severely from the aftereffects.

I like to quote from a person who knew him quite well:

Però il Sammaciccia resta per noi l'aquila che tocca altezze vertiginose, il Maestro sicuro di sé, per cui a volte, pur nella serenità di intesa, richiede una certa tensione per stargli dietro.
Quanto il Marhaba scrive nel suo prezioso volume per esprimere il pensiero, del resto assolutamente cristallino, del nostro comune amico e Maestro, coincide con il mio pensiero in pieno. [1]

(But to us Sammaciccia is like an eagle flying at dizzy heights, the Master sure of himself. At times a certain tension is required to understand his thoughts.
What Marhaba writes in his valuable book to explain the nevertheless crystal-clear thoughts of our common friend and Master, agrees exactly with my own feelings).

This chapter is meant to tell a very unusual story, for the benefit of those who will accept it. But, as Bruno liked to say, it will be like buying a diamond, believing that it is but a piece of glass.

Summing up, this chapter is the result of the work of three different persons, plus many more whom we are going to meet in the following pages.

Stefano Breccia
January the 15th, 2006

Bruno to the Reader

Dear friends, read this book carefully, because it is the evidence of the existence of salvation, both from a spiritual and an human point of view.

I am a very religious person, although not a zealot. I have studied almost all religions, including the most primitive ones, and I feel the need to say that human religiousness is a key to salvation. I am a catholic, but I also esteem other religions. When one prays to almighty God, he is doing good, whatever the name he calls Him. I often live near to my beloved Franciscan friars, because I feel that St. Francis was one of the best proponents of mysticism, of self-negation, of profound wisdom, despite his lack of culture, an example of goodness and love, which he demonstrated both in his life and his body, up to his last minute. That's why I love friar Francesco, and the Franciscans, but I do love also all the friars who, concealed within their silence, perform many more good works than what others perform in the open.

As for this book, read it carefully. For a while, forget what you have learnt up to now, forget what idiots, ignorants and liars say, because their statements are an obstacle, and please do concentrate your mind on what is being said.

That's what I was hoping, while writing these lines.

Bruno Sammaciccia
Chieti, May 2000

Hans to the Reader

When my friend Stefano told me about Bruno Sammaciccia's decision to write down part of his memories, I decided to cooperate, adding my experiences to his, even if I had never met him. I had known of him only by name (we met only after this operation).

I am obviously grateful to Stefano for his work of coordination and enrichment. Reading Bruno's notes, and comparing his tales to my experiences, I should say that in Germany our Friends made us live more technically oriented experiences than in Italy. I am an engineer, like Stefano, and have worked in the TLC field (that's how I had met him), and I believe to have profited from what our Friends told me about their technologies. An important invention in the TLC field has certainly been inspired by them (for obvious reasons, I can't reveal which one).

Now that I have met him, I may state that Bruno is an exquisite man, with a deep culture in morality and spirituality. He was not so well aware of what was going on outside Italy, and so he has read carefully my notes. Our meeting at Stefano's became a session of questions and answers, and Stefano has included in his text what emerged that day. Unfortunately, I do not understand Italian, therefore I've been able to read only the first version of this chapter, the English one.

In Germany we also felt that something grand was taking place, and we too tried to conceal everything, except for a few dubious reports.

We were told that Italy was to become the leader nation on this subject, thanks to the centre of studies that Bruno had built, and when everything went haywire in that country, we felt a deep disheartening and sadness. The Friends told us that a moment of rest and re-thinking was necessary, and that we were to meet again in the first years of the new century, but up to now they haven't appeared.

Obviously we were aware of similar events taking place in other countries in Europe, but only at times were we in touch with foreigners involved with Amicizia. Here too, now, very few are still alive. Therefore I agree with Bruno in leaving these memories for the future, when our Friends return, if they wish to. Unlike Bruno, I have asked Stefano not to reveal my family name, and not to include my curriculum in his book (my curriculum is rather similar to Stefano's, with perhaps fewer international experiences, but similar working environment).

In closing, I suggest that the reader examine carefully the following pages, because, I hope, one day he will be able to live similar experiences.

Hans
Chieti, May 2000

The People Within Amicizia

(Author: let me make a short introduction to what follows: Bruno is going to relate some of his experiences with some alleged extraterrestrials; they said they belonged to a kind of Federation of many different planets. Bruno gave them the collective name of "W56". They have been on our planet for millennia. In Italy, they had built a huge underground base, very deep, and really very large, extending from the middle of Adriatico sea westward to central Italy, and ranging from Ortona northward up to Rimini; the ceiling was 300 meters high, and the global volume was so great that it often rained inside! Of course there were many other small bases, nearer to the surface. These people were hostile towards another race, whose "people" were a kind of biological robots; Bruno had named CTR's these latter. Now, I let Bruno speak.)

In April 1956, Giulio, an engineer, Giancarlo, an accountant, and I, were returning home from a long stroll on the beach, and were talking about a rather unusual discovery, a parchment that we had been given by a friend of ours, who had found it in the bottom of a trunk owned by an ancestor. This parchment was worn out, and almost illegible, written in red and black ink, and on it there was a kind of map, showing a castle in Ascoli Piceno. The castle was supposed to conceal a treasure, or something like that, but it was not clear what kind of a treasure it was to be, whether diamonds, gold, or whatever. We got to my house, in Genova street [2], and went to my study. The room was totally green, because of a lamp, and of a writing-desk, whose surface was made of a single green crystal (green is the color I like best). I experienced a strange sensation, a very beneficial one. In fact, all the three of us were feeling as though we were full of energy, bright, and euphoric, feeling healthy and youth. It was a novelty to us, and we were speaking among ourselves, wondering about this novel feeling. We were sitting in front of a large window, through which we could admire a very wide panorama, and a timber warehouse; the light was coming from our left, with the green lamp giving a feeling of peace. My engineer friend was smoking one cigarette after another, and I told him to stop smoking. At the same time also Giancarlo was getting very nervous.

We looked again at the map, which was at this moment almost useless. It was lying on the table. Suddenly from its place in a pen-holder, one of my pens jumped out by itself, and landed on the map. At first we feared that it might have made a blot on the map, but this was not the case. Of course, we then started discussing what had just happened.

While we were looking at the map, I glanced at the sky, through the window, and said to my friends "How wonderful the sky is to-night". Giulio answered "As always", and, while joking, we decided to get back to the problem of the pen. Giulio asked me "How could the pen have jumped out?". I couldn't answer, although I was the only one in the group with experiences in paranormal phenomena; I had already acquired wide experiences in the field, I had met many masters from the east, and was practicing yoga on a regular basis. I had written two books on paranormal subjects, one about the cutaneous plaques, which was suggested to me by the teachings of professor Giuseppe Calligaris. The other book was on the extra-psychical phenomena, telekinesis, exoteric subjects, and the like.

I took the map and the pen, and also took a mill-board that I used to write on, also bordered in green, which I covered with a very thin cloth. I placed this board over my legs, put a sheet of paper on it, while my friends were holding the map. I took the pen, and positioned it over the board. The pen started to move by itself, while I was just holding it between my fingers. It was writing, perfectly, in Italian. After having written some text in Italian, it replicated this text in an archaic language, then in Latin, then in Greek; then it started writing hieroglyphs, ideograms, and other signs that we were could not understand, but which contained a great sense of harmony throughout. Just after that, in a perfect Italian, it wrote: "Now I am going to explain to you who I am, where I come from, and what I want to ask you; I am here to give you the good and the knowledge".

The sheet was full, I gave it to my friends, and they were trembling; I was not allowed to tremble, because I still had the pen in my hand.

We remained in my room for an hour, or an hour and a half; when we left, it was about half past midnight. It was difficult to part from one another after what had happened, so I had to go with them. We stopped at the crossing between Genova and Firenze street, promising that we were to meet again next evening. (this happened in April the 6th, 1956 – Author).

When we were again together, I went to sit behind my desk, and took the mysterious map from a drawer. We looked at it in silence, everyone thinking by himself. We realized that we had to go there, which we decided to do the following Sunday. We also decided to take with us some additional underclothing, because we figured that we had to walk far into the country, which would cause us to perspire, and therefore we thought we could get to a public lavatory to change our undergarments.

We left on Sunday at 8 a.m. in Giulio's FIAT 600, and arrived at our destination some minutes after ten o' clock. I had already seen this castle, during the war, because I had been evacuated to Ascoli Piceno, and in those days I used to go to this castle by walking through an alley which led uphill, surrounded by pines and spruces. The first time I saw it, I had felt astonished and almost enchanted. This second time, the three of us remained still some minutes, without speaking. Then Giulio broke the silence, saying "We have to start, or otherwise we will remain here enchanted". So we started walking around, looking at the structures, and comparing them with the drawings on the map. After an hour, we had obtained no results, but we still had a strong desire to try to get to that treasure.

Two hours and a half later, we went back to the car, having decided to return to the castle two days later. We went to a coffee-shop to drink something, because we were perspiring. During our explorations, Giancarlo had fallen into a pit, and was covered with dust. We asked the bartender where we could clean up, and he directed us to a narrow street going from the town-hall to Piazza del Popolo, where there was a public lavatory. We washed there, and, as it was some minutes past noon, we decide to have a lunch. After that, we returned to Piazza Meletti, where we had a small glass of Anisetta (Author: it's a local spirit, invented by a Mr. Meletti many years ago), and we then drove back home discussing the day's actions, always aware of this sense of energy, a beneficial and strange effect. We also felt we were being watched, and often we asked each other "Who is spying on us?".

Monday night, we met again at my place at half past eight to discuss what had happened, and to make our next. Although we had slept just a few hours, our bodies were full of energy; Giancarlo had his hair still wet from a shower; Giulio said "We must win the match".

We were astonished because, after a hard day, with all kinds of problems, and having slept just a few hours, we were feeeling very efficient, and always accompanied by these euphoric sensation of well-being, and health. It later happened, that whenever we were with our friends, either in their presence, or under remote connection, we always had such beautiful feelings, together with pleasant scents of pine and incense, and flowers. They were situations not to be mistaken for phenomena deriving from holiness: holiness does actually exist, but it's a different thing.

At the meeting, we resumed our discussions, and strange and different ideas emerged. We were convinced of what had happened, but we were not actually sure what the causes were. Why not an origin closer to us, a more human one? Giancarlo had accepted everything from the very beginning; in that he was probably better than Giulio and myself. From my own part, I felt that it was true, I was sure it was gold, but in the meantime I felt I needed to go to a goldsmith to have my "gold" verified. Giulio was concerned more with technical problems. He was an engineer involved in hydraulic activities, and was wondering whether he could get new solutions from our friends.

During the discussion, someone remembered that in Germany one of his relatives had shown him a book which he read, claiming that after the war the Nazis had continued their researches in biology and genetics, that some beings had been born during these activities, and were living in deep caverns, as big as cities, and so on. Of course that's science fiction, with political overtones, written with the deliberate goal of confounding human minds. However, perhaps, our correspondents were these German achievements? Or maybe they were Russians, from some unknown success of the Soviet research? Who knows?

The next day we went back to Ascoli Piceno. During the trip, the three of us were silent, which was not a good idea, for, when you stay silent, that means that you are thinking, but do not want to let the others know what you are thinking, and that's bad. I was trying to understand what was going on, I asked them if everything was all right, and they said that yes, everything was OK. Giulio said that he was perplexed, because he was spending all of his time thinking about this situation, and "you know, I have a job, I have responsibilities, and if I get distracted who knows what may happen". Giancarlo asked us in a rough manner how could we actually be convinced, and I was trying to remain calm. When we reached Ascoli Piceno, the debate was over, they became quiet, and by then we were as usual in the central Piazza del Popolo. On one side of it there is the ancient cathedral of St. Francis of Assisi, and nearby the town-hall, a monument from the Middle Ages. We took a short walk around, just like prisoners, without speaking. Then I told them that it would be better to go to a bar to have an Anisetta, and the other two agreed, Giancarlo sooner than Giulio. Giancarlo had always loved a lot to eat and drink. While we were sitting, we suddenly began to feel that we were being called. It was a very strong sensation. So we went to the castle, and there, at the top of the road, I saw my friends calmer than before, really relaxed, for we were close to the origin of the call, and surely that's where it came from.

All of a sudden we saw some specks of light moving in the air, something like the small seeds that come flying all over in spring, or like cobwebs, but in this case there was light within them. We heard a voice, coming from nowhere, a very calm and strong one "Now, my friends, stay calm, because I am going to have one of us appear. Are you ready, or aren't you?".

Giulio was very upset: "What happens if someone comes along and sees what is going on?"

"Be sure that while our friends are with you, nobody else will be allowed to intrude; if they do we will divert them away".

A man came out from behind the wall, followed by another. It was our first encounter. We saw these two persons emerging from a narrow track encircling the castle. We were very upset. One of the men was very tall, the other very short. We were just in front of the main entrance, and they came towards us, speaking our language perfectly. At first, in the dark, we thought it was just a joke. But, as they approached, we saw that one of them was more than 2.5 meters tall (Sinas was the name he had chosen for himself), and the other was about one meter tall (his name would be Sajù). The latter had a squeaky voice, as dwarfs often do, but his body was perfect, although smaller than usual. His voice was that of a powerful man, a man in charge.

They came to us, and we felt an intense emotion. They both shook hands with us, very gently, because we felt that they were very strong. We felt at that moment a strong sense of love coming through their hands.

We were totally calm, even to the point of making jokes. Giulio asked "Are your women also as tall as yout?". He understood immediately that his question was out of place, and excused himself. It was his self-conscious reaction to what had just happened.

We remained there for over one hour and a half, discussing about everything. At one point, I told them: "Look, my neck is hurts, because I have to force my head back, so that I can see you. Let's sit down". He agreed. There were some steps nearby, surrounded by three small trees (which have now become really big - Author), so we sat on these steps, the smallest one on the top, the three of us in the middle, and the tallest one some steps down. How many things they told us! That theirs was an important mission, that they had been here for many years, that he had been here three times, and that three or four centuries ago he had been in Central America, because in that area there were bases operated by other aliens, which he had helped to get hold of, and that there was a war, unknown to us. He said that they usually kept to desolated areas, where nobody could see them, so that they would not bother anybody, and at the same time they could not be bothered by anyone.

He said that usually they get mimetic, so that nobody can see them, and I asked "Why, you have been seen at times!", and he answered "It must have been when we had diverted our attention. We are beings just like you, after all".

They were perfectly aware of our history, our religions, our philosophies. The very first words they said were "This Earth was made for the good, but the men who inhabit it are transforming everything into bad. We are not here to conquer, we have nothing to conquer; our interests arise from the fact that your Earth lies within our stars, and so we are concerned with it. I do not live on a planet, but everywhere I happen to travel".

"This is a critical point in your history, a turning point in your technologies, but because of your childish enthusiasms you are forgetting your moral values. That would be a pity if you forget them, because everything arises from morality, and everything is done because of it. For this reason, we had, and we are still having, many problems with your people in the Middle East, and you too are going to be in trouble with them in the near future." [3]

He was speaking a perfect Italian, while the other one was not so good at it; but their subjects were different from our usual ones. We speak about politics, eager to achieve a social goal. We want to emerge, to do something in order to get new means, while to them all of that was nothing. They were speaking of morality, and that they were living according to that principle. They said "We are going to receive a boomerang (they used the very word "boomerang") from what we have done, that will allow us to live as well as you are seeking to, do you understand? Our good and our reality will be stronger than your doubts".

Giulio, who was a bit materialistic, said "If I do not work from 6 a.m. to 8 p.m., I will not make enough money to live". He answered: "This conditions of yours is due to the fact that also work has became an unpleasing thing for you: working does not satisfy your expectations. We are satisfied with what we do."

A long time had elapsed. "Now we must get back", he said. I had a watch with a luminescent dial, and it was three a.m. We parted warmly. "We are going to meet again, from time to time" they said, "but probably in a different place. We'll let you know about that".

We did not want to let them go, so we started asking them to stay longer with us. He turned himself and said "I know that you still have many questions, many doubts; this attitude is a correct one. But we will try to answer your questions later and to remove your doubts, and also to teach you something very important". Then he embraced each of us; I felt a strong warmth, a good warmth, like when you are cold, and cozy up to a fire place. The small one came over to me; I bent down, and he put his hands on my shoulders. Then, he pressed my wrists, and left.

We stayed there until sunrise, discussing what had happened. When the sun rose, we realized what time it was, and Giulio got up in a hurry, because if was time for him to go to work. He worked near Popoli (some 40 km West of Pescara), so he got the two of us to Pescara, and then he went to his job, without even stopping at his home. Giancarlo also had to go to work. He said "I'll take a shower, then I'll go to work; but how can I go to work, after all that?". Actually we were not tired, but only over-excited. We would have liked to have more time for discussion, but it was not possible, so we all went about our usual activities.

That night we met again, at 9 p.m., with the usual discussions. They had told us to go to Colle Orlando, a small hill to the South of Pescara. In these days, an engineer, a friend of mine, was making surveys there. He had told us that they were going to cut a road through the hill, and indeed some time later I noticed that actually they had cut the hill into halves, and in the center there were still machines working on this job; but this was not of any concern to us.

The next afternoon we went to Colle Orlando. We left the car at the end of the road, and went on walking, taking a small transistor radio [4] with us. Suddenly, through the radio they told us to stop and to arrange ourselves into a triangle. At its center we saw a twinkling light, a kind of vertical light. It was like a very long brooch, a kind of a needle some 50 cm long, with three crystal balls at the upper end. They told us: "Put your hands on the spheres", and we obeyed. Giancarlo actually grabbed one of them, as if trying to take it away! They said "Do not throw, just keep your hands on them". The spheres were some three cm in diameter. From the top of that long needle, something like puffs of phosphorescent smoke started to come out. They rose up to the height of our faces, and then disappeared. After a while the brooch of light entered into the ground, and disappeared. Then we were told to go back to the car. That particular place had been transformed into a kind of facility for us. A subterranean base was to be built there so that "… when you come within 30 km of here, you will be able to communicate with us by thought, but only when you are ready to do so". They also told us "Look inside your pockets: you will find small plates." Giancarlo put his hand so strongly into his pocket that he almost poked through it. I too looked inside my pocket, and found a small, rectangular plate, whose color was between platinum, and silver, with a lot of, say, small diamonds all over its surface. They were not real objects embedded into the surface. If you rotated the plate, small specks of light sparkled on its surface. They told us to wrap these plates in a sheet of silver paper, and to protect them very carefully. We decided to keep them inside our wallets. These plates were another method to get in contact with them. "As a third means" they went on telling us "you should put copper plates inside your shoes; copper will also do you good, because it will insulate your body against electric fields". (in the following days, Giancarlo put two actually massive copper slabs inside his shoes). We did what we had been told to do. I went to Di Giacinto's shop, near to the railway station in Pescara, and bought a couple of thin soles, put two small copper plates inside them, and since then I have always used these "copper soles" inside my shoes.

"Now we have to do something to your dog" they went on. Were they going to transform my German shepherd into some kind of electronic dog? As if they had read my thoughts, they told me that there was no problem; on the contrary, the dog would be better off after the operation. But, as a dog has four poles in contact with the ground, while humans have only two, they told us that they were going to make not two positive poles, nor two negative ones, but rather they were going to fit my dog with electrically oscillating poles. "How are you going to do that?" I asked them. "Don't worry; take your dog for a walk along the seashore, as usual, and when we advise you, send your dog into the sea, and make it swim in the water".

Dik, my dog, was a very good swimmer, and was very fast in the water. When the following day came I sent it into the sea. It entered the water, and then, all of a sudden, it looked at me, and made a strange sound, and Giancarlo said "You see, now they've put the gizmos into your dog!". "Now, observe how your dog reacts", they told me. And, actually, my dog came back to the shore, lay down on the sand, and started licking its paws. "Now it has four nuclei inside its pads, that won't hurt it; we could not have done it any harm, because we feel a great respect for animals too. We really love them".

When we had to get in touch with them, we usually arranged ourselves in a circle, with Dik in the middle- The dog usually was lying down, and as the communication began, I felt the process happening much easier than before, it's difficult to describe exactly how. Before I feared that the connection might be broken if I made the slightest move, and I took great care not to alter the situation. Now everything was much easier.

"Beloved brothers, I come from the stars. Up there, far, far away, inhabited planets exist. I do not know why I must do that, but, as it is my duty, let's try to cooperate together, in peace and harmony. Now I'll give you some directions that you must follow exactly. For the moment, let's take a six minutes pause". During these six minutes, we looked at each other. We were very impressed, but in any case we were feeling fine and very happy. Some minutes before resuming the contact with this unknown being, we felt within the room a strong smell, a very pleasing one. After exactly six minutes, it started again, and they told us to go back to the castle, to look at one specific place, at the right side of the main entrance, then to go to the outer wall, and look for a black rock among the bricks, and to place, each of us, our left hand over it for a couple of minutes.

The next day we went again to Ascoli Piceno, and once again the radio in the car switched on by itself, and we were told to tune it, and listen. To our surprise, caused not by fear, but because it was a new and inexplicable fact, a calm, profound voice came out; not the voice of an ogre, but that of a good human being. It made me think of Buddhist monks who, hours after hours, repeat "Om". First one being, than another one, began speaking. They told us a lot of amazing things about human beings, about their great possibilities, what we should know but do not, about their anxieties concerning our civilization, which had taken many wrong paths, forcing natural and terrestrial laws into dangerous reactions, capable of producing great damage and destruction. Now I do not recall every word, but I can state with certainty that those beings, unknown to us, were using words which carried very deep meaning, so that it would have been wrong to consider them fantastic or abstract. They were real words that the men on earth should have understood.

The next contact did not happen at the castle, but at my own home. We had several radio sets, among them a small Geloso [5]. We were told to wash our bodies, and to wash our minds at the same time, thinking of religion, and avoiding thinking of wars, and the like. We entered my room, at 10:30 p.m. It was at the end of April, 1956. The radio set was sitting at the center of the table, and the three of us formed a triangle around it. A voice came out from the radio, saying: "In the near future, it will be easier to us to communicate from mind to mind, and, to do so, I choose the one among you who masters the abilities we use. He will have to go, for an eight-months period, to the top of the hill named Colle del Telegrafo, he and his dog, from 11 p.m. to 6 a.m., rain or snow, hot or cold".

I did so. After the third month they asked me to stop, because they wanted to review what had happened. They put a small nucleus behind my left ear [6]. This nucleus was supposed to activate, inside my mind, the ability to communicate telepathically with them. They said that it was the same apparatus they used when they had to operate instruments from a remote site. When they introduced it into my body, I felt no pain, nor disturbances, nothing; they told me that it was for my own good.

For four months longer I had to go on with this process of climbing Colle del Telegrafo at night, stay there, and return in the morning, always with Dik, my German shepherd. Nothing happened to me, not even did I catch a cold, although in winter it was a very chilly place. I used to take a bottle of water with me, a thermos full of coffee, some aspirin, and during the nights I poured some coffee for Dik. At first, there were many by-standers, who probably figured that I was totally crazy, but after a while they got tired of spying on what I was doing in that isolated environment. I didn't care too much about the people around, but the presence of my dog was always a great help to me.

During this time, and after that, we went on, spending afternoons in long discussions, walking along the sea shore, with all of our friends that were part of our small "stars community". We received so many revelations, so many predictions. They elevated us, and made us understand that there are many differences between true wisdom and that which comes from an unenlightened, presumptuous mind, which usually deals with unknowns through his self-sufficient knowledge. They told us that knowledge may dwell inside simple minds, while false knowledge may inhabit ambitious minds, which have the sense of personality, and the will of dominion and conquest, but not the sense of individuality. Man cannot conquer anything, nor can he create anything. The unenlightened man, usually, has a consciousnee focused on possession, while the man who has a strong sense of individuality seems to show an apparently weak personality. He usually speaks very little, but when he does speak, he has something to say, and often his listeners do not understand what is being said to them.

These lectures, on several subjects, went on day after day. They were obviously not intended to transform us into geniuses. Often they dealt with facts that we, terrestrial men, at our cognitive level, were just able to accept, but not to understand. For instance that an inanimate object is aware, in one way or another, of the feelings of the people around it, and reacts in a quasi-intelligent way to the attitudes of those who try to use it in an inappropriate manner. They spoke of their love for the universe, for the whole of creation. They were always speaking about the blade of grass, saying that the whole universe lies inside a single blade of grass (That's Fractal Analysis *ante litteram*! - Author), saying that if we love a plant, it will grow better, while if we hate it, it will probably die.

One day, they told us to go to Ascoli Piceno, to meet three more of their people, whose names would be Luxor, Siderius and a real giant whom we were to name Romolo [7], who was so strong that he could literally bend a cannon! But he was also of a very gently nature, he was used for certain specific jobs. In short, Romolo was an enlightened person, not at all a porter or something like that.

The three of us went to the castle one afternoon, sometime in the spring of 1957, probably in March, for Dik was just a puppy. Probably it was march. Sigir told us: "Now that we know each other, you understand that we are not phantoms, or devils, nor are we the products of ancient technologies invented by the fools that have ruled your Earth, Hitler and the like". "OK", we told him, "introduce us to these new people". We had no fear, but any way we were very impressed by this strange and unusual situation.

Ask my wife, who entered the kitchen of the flat we had in Milan, and found there, sitting on the floor, Dimpietro, a man more than three meters high, who was sitting so that he would not frighten her with his height. We came to call this man "the poplar", because of his height. He was a true leader, not a domineering one, but still authoritative.

When this happened [8], my wife had just come home after her shopping, found this incredible being seated on the floor of our kitchen, got frightened and ran into our bedroom, locking the door behind her. I had told her about our friends, but she had never met one of them, especially our "poplar". In the meantime I was walking in the neighborhood with my Dik; deciding to return home, I looked for the caretaker to let into the building, but he was nowhere to be seen. So, I rang the bell at the intercom, and my wife opened the door. As I entered my flat, Alessandra, my wife, told me that there was "somebody" in the kitchen; Dik had already gone there himself. When I entered the kitchen, I found Dimpietro seated on the floor, and Dik sitting beside him. My wife, who was still terrified, told me "Take care of your friend", and flew away to the bedroom. Meanwhile, Dimpietro remained seated, without uttering a word. Then he got up, his head just gouching the ceiling. "How will we talk to each other" – I asked him – "with a megaphone?" "That's why I sat on the floor" "So sit down again" I said, which he did. He told me "Your wife is terrified of me, but do I look as though I would terrify anyone?" "It's not that you go around terrifying people. She knows that you're a man from another planet, and she's very upset, do you understand me?"

Dimpietro was silent for a short while, and then he told me "Look, now I have to smoke" In those days, I smoked just cigarettes, and I offered him my packet, but he turned them down "No, these are just for children. You go downstairs. You'll find a blue Citroen, with a white roof, parked just in front of your building; here are the keys; do not get upset when you see that there's no driver's seat, because I need to sit directly on the floor of the car. You'll find some cigars inside the gloves compartment" I did as I was told to, found four cigar boxes, and came back with one of them. Dimpietro opened the box, took a cigar, then he asked me where the could leave the cellophane. I didn't answer, so he threw everything into the sink. He told me "Call your wife in. We have to calm her down and convince her that I'm not aggressive, and that I don't cannibalize other people". I went to the bedroom, and found my wife pale and afraid. I asked her to come back to the kitchen, and to make some coffee, because our guest was fond of it. She was still troubled, and had problems even with making coffee, as her hands were trembling, but she managed to finish. She had a "napoletana" (Author: it's an old kind of coffee-pot, from Naples, that you had to turn upside down at the right moment; I barely remember having seen one when I was very young). "That's the coffee-pot they use in Napoli" Dimpietro said; "You know everything" I answered him. "There are three kinds of it – Dimpietro went on – take care because you must reverse it as soon as the water starts to boil, but be sure to reverse it in a quiet way". My wife was looking at me rather confused. Then, she took some small cups (Author: as many Americans are already aware, Italian coffee is very concentrated, and it is served in small cups), but Dimpietro protested "No, I need a larger cup!" So my wife served him the coffee in a milk cup (Author: about a quarter of a liter) which means that he got almost all of the coffee. "Forgive me, but I've gotten used to coffee" "Do you need any sugar?" "No, no sugar at all".

In the meantime Dik was looking at him, at times annoyed by a fly, but always very attentive. Later, as dinner time approached; my wife asked me "What should I make for your guest?" "Make me – he answered – a "frittata" (Author: a kind of an omelet, Italian style) with hot peppers" "Have you seen a pepper plant?" "No, I haven't, but I know you come from Abruzzo (Author: the Italian region with Pescara, Montesilvano, and the other places already mentioned), and people there are used to peppers; and, please, do not spare peppers" In the end, he prepared his "frittata" by himself. We offered him some bread, but he refused, because he said he wasn't yet accustomed to our bread. Instead, he asked for some wine. In my kitchen there was just some white wine, and I knew that he drank only red wine, so I phoned to a nearby grocer, and in a few minutes we had a bottle of "Corvo di Salaparuta" (Author: it's a very good Sicilian red wine).

I opened the bottle, and Dimpietro said "Would you allow me to drink directly from the bottle? I am used to doing it this way. If you like, I'll pour some wine into a couple of glasses for you, and then I'll drink from the bottle" And he drank the whole bottle, in three mouthfuls. "Doesn't that hurt you?" "No, you must understand that it is not the quantity that hurts, but the quality" "And don't you feel heartburn, because of the hot peppers" "Yes, but I like it; moreover, as our intestines are longer than yours (mine in particular!) peppers help peristalsis, and that's good, because I am able to defecate on a regular basis. That's usually a problem for our people".

Then he smoked again, and the time came for him to leave. He knelt on the floor, embraced my wife with great delicacy (she was still a bit upset because of that unusual dinner), and told her "Remember, I do not eat women, I only eat peppers, pasta, and some sweets at times; the next time I'll come, I'll let you know in advance, so that you can prepare some sweets for me". He kissed her on her forehead. It was a strange vision, I can assure you, looking at this extremely tall man, with a stern aspect, kneeling on the floor and trying to calm down my poor wife.

It was three o' clock in the morning, and he had to drive the Citroen parked in the street. Luckily it was night, and nobody was around; of course he could not use the elevators, as he simply could not enter it; so Dimpietro, Dik and I went downstairs, with a great care. He opened the car, and I could see that there was no driver's seat, so that he could sit directly on the floor. He entered just as if going to bed: first the legs, his feet in front of the right seat, helping himself with his hands, then sitting on the floor and finally forcing his legs on both sides of the steering wheel. "Do you have any problem with the pedals" I asked him "Not at all; if necessary, I can use my hands to operate them". Of course, this was true.

He started the engine and began to move. I asked him "Do you know the way?" "I know every street, even the alleys" he answered, and sped away. I was feeling as if my own father, or mother had left, a very strong emotion.

Let's return to that March afternoon in 1957. When we arrived there, we were told to wait until night, because the place was rather full of people. "In any case it is good that you have come, because at night you will see some ships you haven't seen before". "Fine, but why didn't you tell us in advance, so that I could have brought my camera with me?" "You would not be able to take a picture of them". "Why not, if I can see them?" "It's a different situation. There are differences that you cannot understand". In the late afternoon, we were relaxed, waiting and smoking. They had told us not to worry, they were not going to appear all of a sudden, like aggressors. They would tell us in advance what would happen.

We asked the permission to go downtown to find something to eat, so we walked down and entered a café. We saw a large tray full of freshly cooked candies, to which Giancarlo and Giulio paid the honor it deserved. We were very merry, almost euphoric. I drank a small glass of Anisetta, a drink I like very much, and then, while my friends were still eating, I went out and entered a near-by church. I was feeling gloomy and happy at the same time, an unexplainable sensation. When I finished with my prayers, I went back and looked around a bit. There was a magnificent medieval arcade, and nearby a printing-house, and a bookseller. Looking at his shop-window, I noticed a book about St. Ignatius of Loyola, and as I had long been looking for information about him, I entered the shop, and bought the book, then went back into the café, only to find my friends still eating!

By this time it was nightfall, so we climbed uphill again to the castle. Some twenty minutes later, they told us to follow a narrow path, and to go down some steps, into a woods. Looking at the wall on my side, I noticed a yellowish foam coming from the cracks between some bricks. I was told that it was a sulfurous substance, slightly radioactive; "When you pass by it on your way back, shine your flashlight on it and you'll see that it is also a bit phosphorescent".

We were met by the three newcomers; they approached us, smiling. They were not exactly beautiful, but we could not say they were ugly, either. The first thing Romolo told me, in a perfect Italian, was "Our dear friends, what were you expecting? To see monsters? Look how handsome we are"; Giancarlo, as usual, made one of his little jokes: "Yeah, but, as big as you are, where can you be seen in public? It's better to be little, may be even ugly". "Look, we are not here to bring you the civilization, but because we are following a plan that has been designed by someone higher up than us, in order to study and to intervene, and to allow you to understand us in the proper way". "Do you know me now? Pay attention to me, and remember that you will have to do what we tell you. If you do not obey us, you are going to hurt yourselves. You have been lucky. Be quiet and you won't be hurt, even if our enemies try to create ill will among you, and do things that you can't even imagine. We are aware of that, and so it will be up to us to protect you."

Dimpietro had rented a small isolated house in the country, near to Forlimpopoli [(9)], which was owned by an elderly woman. He lived downstairs, while she had retired to the upper floor. He was used to cooking his own meals. He told the woman "Look, I really enjoy playing the violin at night". And she answered "Do play whenever you like, there is nobody around". Indeed there was no building around, just a narrow path, 1.5 km long, which lead to this house. And so, at night, he would go out, and play his violin. Think how small the violin would be for a 3.5 meters tall man!

When I went there, he told me that he was waiting for an Austrian, or perhaps a German, a Mr. Gustav, "who has to come, because now he must now go underground". He asked me to find a way so that the both of them could go to Ascoli Piceno, where the main entrance to their big underground base was. I had to rent a large car, remove the front right seat, so that he could sit directly on its floor, thus appearing to be a normal sized man sitting in the car. I asked him why there wasn't a tunnel from his house down to the base, and he answered "Why? I can enter from near the Rocca Pia in Ascoli Piceno, so there is no need to dig so long a hole; I am not a caterpillar!".

So we arranged to take him to Ascoli, and arrived there at 2 a.m. We went to the castle, and found three of them waiting for us, Sigir, Meredir, and another one whose name no longer remember. We shook hands, and then Dimpietro said "Now I must go down". He spoke perfect Italian, and moreover he knew most of the Italian dialects. At times he made us laugh, telling jokes in a very serious style, which added to the fun. He went down. We stayed outside with Meredir, talking about so many things. He told us that we would meet their younger people, for there were youngsters with them, and they were having difficulties adapting to their situation. It was a sacrifice for them to remain hidden underground to work, taking care not to be seen around, with the CTR's [10] who were attempting to make damages.

One afternoon, Giancarlo phoned me saying that that night he would be a little late in coming to my house. "If there is anything important – he added – please call my house number, because I will be there this afternoon at home, seeing a person to discuss problems with my new house".

Then I got a message from our friends: that night three spaceships were to arrive, and the three of us (Giancarlo, Giulio and I) were asked to be at Ascoli Piceno, for our presence was necessary at the event. I tried to phone Giancarlo to inform him, but nobody answered my calls. Later I learned that a general telephone blackout had taken place in Giancarlo's area. Therefore I got in touch with Gallarate [11], and he told me not to worry, as he would solve the problem. He told me to dials a very strange number on my telephone. It was a very long number, with some thirty figures in it. I did so, and this time Giancarlo actually answered my call. He was amazed to hear his telephone ringing, because he had been told that the breakdown would last until the following day.

So that night the three of us got together, drank some tea and went to Ascoli, arriving just in time. We stayed outside the Rocca Pia, waiting for something that none of us could describe. All of a sudden, the sky changed. It was as if the stars were painted on a solid background, but not a steady one. It was more like a veil, that was being shaken as if a strong wind was blowing. But in reality there was no wind at all, and usually wind doesn't shake the sky!

"What is that – asked Giancarlo – a fishing net?" "Don't be silly" said Giulio. Far away we saw three very small little spots, approaching the place; they were very shiny but their light was not blinding us. The ground under our feet started to tremble, so strongly that Giancarlo was thrown off balance, and fell down. We were astonished. It lasted some 15 minutes, during which Giulio took shelter inside the car, and

Giancarlo sat on the ground. "Ask them what's going on" he told me. They were almost hysterical, and I was not able to quiet them down.

All of a sudden, two of the lights grew larger and disappeared. We realized that they were the spaceships we had been waiting for, and that two of them had already entered the underground base. The third spot just switched itself off. A few moments later, Gallarate, Sinas, and another alien appeared, together with Dimpietro; "What's all this mess, what are you afraid of?" he asked us in his strong voice. Then he started to pull our leg: "I didn't bring any toilet paper with me - I didn't expect you would need any". We were happy being with them, but at the same time we were feeling a bit uneasy, because one could never tell what was to happen when Dimpietro was around, a notorious practical joker; this time, though, nothing too strange happened. All the six of us sat on the ground. Dimpietro took a big cigar out of a box. He threw the empty box away, admonishing us to pick it up before leaving. Then he broke the cigar into four parts, keeping one for himself and giving us the other pieces. Then, he lit his cigar, with a flame coming out of … his forefinger, and laughing at us!

In the meantime, I noticed some strange activity. Some aliens were coming toward us. They were walking past, disappearing behind the wall. I was a bit curious about what was going on, and began to follow them, but was stopped by Dimpietro: "Where do you suppose they are going? At home we have several houses, but here on your Earth we have only this place. They are entering our base" "But I can't see them going in" "Well, but you know how we are - we like to be a bit spectacular, at times". Dimpietro, at his most typical.

I went on: "Does the door close after each one of them?" "No, it doesn't" "Then, may I go and have a look?" "You're welcome to". Then he stood up, and, with his great height and size, he lifted me off the ground with one arm, and Giancarlo with the other. As I was wearing suspenders, they broke off, so I was going to lose my trousers along the way. When we got to the entry, I saw an opening in the ground, like a vertical tunnel heading downward. I thought that the tunnel might have weakened the castle foundations, and, as if reading my thoughts,

Dimpietro said "Do you believe that we are such fools? We have taken care of strengthening the structures, so that there is no risk". During all that, the procession of people was continuing; they were carrying parcels of every shapes and sizes. Everything was wrapped in a kind of silver paper, or at least it looked like that.

Dimpietro took the two of us back, and said that he had to go down in the next ten minutes, so he urged us to light our cigars. We refused, and I kept my cigar as a souvenir. But after so many years, it dried out and crumbled, and now there is nothing left of it.

Then we left. As usual, Dimpietro embraced each one of us, which meant that he had to kneel down so that his arms were more or less at the right height. We approached him one at a time, and embraced him. He was saying "Please, let the world know that we have come here with a great love towards you. You speak about love, but you do not know what love is. It is the very basis of life itself".

We went downhill, to the usual Meletti coffee-house, and stayed one hour longer, discussing what had taken place. It was now three o' clock in the morning; the waiters had already gone, and there was only the owner of the shop, an old acquaintance of mine, who excused himself, saying it was already long past closing time. Thinking back to those times, I remember that we were actually always very close to one another, always surrounded by a feeling of well-being and elation. When we were going to meet our friends we felt like young boys on their first date with a girl. We were happy, excited, and nothing could have prevented us from being at appointed rendezvous at the appointed time.

We got back to Pescara and stayed talking in front of my house, for another half hour. We would have liked to stay there the whole night long, but Giulio had to workin the morning, and needed at least a little rest.

In another occasion Meredir [12] told me to go to the Rocca Pia, taking with me my tape recorder; it was a small Grunding, battery-powered, metallic unit, an office instrument. I was told to lay it down on the ground, and to start recording. When later on I listened to the tape, I could hear clangs, creaks, cries, shouts, because, I was told, I had in fact recorded the noises of an actual battle that had been raging inside their base. They had been able to seize their enemies; both of them had powerful weapons, but our friends had also a conscience, while their enemies were cool, hard people.

Then, at last, we were told that we were allowed inside their base! Giancarlo and I were in front of the castle, and were told to wait; in that area the day before Giulio had left his car, a blue FIAT 600, and had left it there for our use. He should have reached us, but was not there, and we were wandering why; only some days later he told us that he had mistaken the day.

Then, we were told to go to the right side of Rocca Pia, and to stop at a certain point of the path. I started feeling that the ground under my feet was kind of trembling, the same sensation you feel when you stay near to a pneumatic drill in action; I was fearing that may be there was an empty room under us, and that the ground was going to collapse into it, because of our weight. On the contrary, the ground opened itself, and somebody came out. We were unable to speak and that man told us to get down with him, but I couldn't ever understand how. He told us to proceed toward the empty area in the center of the hole through which he had ascended; I was fearing I would fall down into it, but he told us to place our feet in certain areas (there was nothing at all visible in there); I did so, and felt as some invisible step was preventing me from falling into the pit. Then this invisible floor started lowering into the vertical corridor. When next day we got out, I could see that my feet had grown slightly red, and later on I was explained that it had been because of the process of descending into the base and of ascending out of it.

The descent stopped inside a huge subway, with crystal-like walls. It was filled with a soft light; we looked around for the lamps, but were told that there were no lamps of any kind. "You cannot understand – he told us – but this place is filled with a peculiar radiation that interacts with the energy of the photons, it's a bit as if we could switch on the photons; moreover, they get continuously re-charged, while this radiation is active". The light was beautiful, a light pale blue. We could also feel a scent, and the air was very clear, we could see far away. It was strange; in later times I was able to verify that, even if you were smoking, you could see through the smoke, with exceptional detail. There were no shadows. "Look – I told Giancarlo – there are no shadows". "I know, I too had invented something like that, years ago", and our friend smiled.

Our escort was Meredir; almost immediately we were met by Sinas and another one, whose name now I do not remember. We walked with them, for more than ten minutes; it was a pleasant feeling, walking with these three friends of ours, inside that huge structure. I was feeling calm, with a well-being I had never felt before. It was as I would receive doses of well-being as I was breathing the air. They explained that the air was different from the one available in our towns: it was full of negative ions, that were the cause of that sensation. "Touch your hair". I put an hand over my head, and felt that my hair was stiff and creaking, as if frozen. "You are being detoxicated, when you'll get out you'll feel better."

"But now let's talk about more important questions: this is our environment, where we live; in that direction there are the youngsters; in this moment they are inside a kind of class room, where they are studying." "May we see them?" "Yes, but secretly; we do not want to frighten them, because what they know about terrestrials is no good news. They think of you as if some kind of wild beasts."

We went near, quietly. He pushed a kind of a button, and a small screen appeared over the outside wall, a white square 50 cm in side, and an image appeared over it; I was able to see inside the room; it was some 50 meters long, I cannot say how wide, because I could not see its sides. It was as if there was a video camera in action inside the room, a camera that could change its bearings, and could move by itself. This way I was being shown the so-called youngsters. "How tall are they?" "Two meters thirty, two meters forty, even two and a half." "And they are young!" "For us they are boys; someone is 15, some is 30, another one is 95 years old. Biological growth to us is slower than yours, but achievements are quicker.". I saw that some of them were wearing a big cap, similar to the drier our women use while at the hair-stylist, and asked what was the use of it. "It's used to increase mental capabilities, but not in an artificial way, we would never do such a thing; rather, it gives a benevolent solicitation to their nervous system, and in the mean time it detoxicates them. Intoxication prevents the full evolution of men; would you remain fasting, or would you feed in a different way (unfortunately we behave largely as you do), things would be better."

Many of them had short hair, as our German boys, a stiff hair; they were showing a benevolent countenance; some had brown eyes, other very light colored, green, blue eyes. They were of different races, and I was told that there are actually many different people, but that in most cases only their morphologies are different, not their biologic functions, and that, of course, differences exist also among men belonging to the same race.

"So these boys are already at a scientist level." "Yes, but we take care much more of moral aspects, because they get in touch with powerful weapons, and if they were not provided with a strong ethical attitude, they could make great damages. This same ethical attitude we have imposed to our instruments, even to our weapons; if you would try to use them to do harm, they would not work, or even they would disintegrate themselves."

We walked further on; I was thinking of what we had been told, of their willingness to act only for the good, and was feeling compelled to let everybody know about them; Sinas, who was leaning against a kind of column, told me "No, you cannot do what you think; do not let what you have seen known outside, you would engage in very distasteful situations." "I haven't told you anything" "Yes, but you have thought about it!"

Later on, we were offered a drink, a very pleasant one, like a lemon-squash in color, but with a different taste. "It is not lemon, nor is it synthetic; it's a squash of our fruits, and it does much good; men and animals may drink it, and even vegetables; if a plant is not so good, just pour some drops of this drink on it, and you'll see it recover quickly." I said "Fine, let me take away some of it." "No, out of here it doesn't work, because this drink has the virtue of detoxicating, without any side effect, but it gets quickly intoxicated itself by external agents."

Later on we entered a large, circular room; on the top of it, something like a carousel: lights of every color and shape were revolving around over its ceiling. "Well, we are just cleaning up. Those lights are our cleaning operators. Don't worry, now I'll switch them off." He put his hand inside his pocket, and the lights disappeared; now the ceiling was a single, compact, crystal mass. I can't say, because it was not transparent, nor was it opaque; it was neither. We remained there, speaking with persons that I didn't know, then Sajù appeared far away. He was like chili pepper, he was to be found everywhere (Author: this remark refers to the Italian habit of making a large use of chili pepper in cooking). I heard his crispy voice, I asked "Is there Sajù?", and "Yes, he is everywhere", I was answered.

I do not remember if I had already taken Dimpietro to this base from Forlimpopoli, probably not. By the way, did I tell you about the car that was running without touching the road with its wheels? The car that was able to take off, fly for a while, then land again unto the road. One time we went to take Dimpietro with that peculiar car.

Probably Dimpietro was not inside the base, that day, because I remember that our friends were telling that he was to come within some days. "When Dimpietro will come – they were saying – our organization will start to operate, in the way it has been stated. Do you see that machine up there?" It looked like one of our electronic device, full of buttons, lights, screens and the like. "No – he said – it is not an electronic device; it is very different from yours. It is not a file storage, nor a memory. If you look inside it, you'll see it's empty, but it contains a load of energy that will suffice to us during a whole year.". He opened its cover, and inside there was a screen, and over it a wide light was moving, without making any noise; it was a light, a dark green one, but it was as if there was some matter in it, may be one could even touch it. It was like a boiling broth. "This is energy at its initial state, it may be transformed into solid energy, or into even more subtle energy. And this depends upon this small instrument near by." He showed me a small circular knob, with light marks on its periphery. I was not able to understand what it was. "You see, if you touch that knob in some point, it will select the kind and the amount of energy you require, then it will distribute it wherever you like, inside this base." "How can it be done?" "Look" He opened a door, and a kind of a map was there. "If we have to send a certain amount to energy towards a certain place, I select the destination this way." In some way he selected a place, and a very strong bluish line appeared on the screen, the it disappeared. "I have charged with energy that place, and it will be enough for seven, eight of your days."

"But, tell me, do you eat?" We had asked this question other times, but had not understood their answers. "Of course we do eat – he told me this time – only, usually, we do not eat your kind of meals." He took me inside another room, where there were many stools, around a table, and an altar on the far wall; do you remember the refectories of friars? Actually this room was different, but had this kind of appearance, because I was told that our friends collect into this room to have lunch, or dinner; they sit around this table, and eat without speaking. Giancarlo usually would speak a lot while eating, so I told him "Giancarlo, you cannot sit here, because they will throw you away as soon as you utter anything." They at first sit there, and pray. They are used to pray, they pray in full sincerity; their prayers are not just a mystic instrument, they are able to get energy from this activity. Then they all look at a specific point on the altar, and this point starts to light up, and they wait until this point reaches its maximum of light. They told me that during this process, this point grows also in dimensions, and at a certain moment it develops a ring, like Saturn, then this ring starts to pulsate, and that's the sign that it's enough. "For us it is a good practice, because while we are praying, and filling the environment with our psychic energy, we wake up other forces, and in the meantime we awake also ourselves; in those moments we are like drunken people. Energy may hurt, but to us it is good."

We had arrived to the end of the base. Now I remember that Dimpietro had already been there, and had got out. I remember that, because, in a small room, I saw a Moka (Author: it is a typical coffee-pot brand, widely spread in Italy), and they told me that Dimpietro was very fond of using it to prepare his own coffee. He liked very much Italian coffee, but also Italian food, wines and spirits, may be even too much. In his room there was this Moka, a set of small coffee cups, and many other things typical of our environment, because, I was to learn, Dimpietro liked very much the Italian way of life. In his cubicle, I saw also a very little plant inside a glass, without any water, and I told my friend that that small plant was going to die. "Touch it". I did, and a great many tiny specks of light aroused around this little being. "It may live this way for years." These very little things were playing an important role in their civilization; moreover, there was much of our culture in their environment, and their technique was never deprived of human factors. They put their moral in front of their technique, while here we do just the opposite.

Then we got out of the base; it was 3 a. m., and over our heads there was a very clear sky full of stars. I exclaimed to Giancarlo: "Look, stars are still there!". "Why, what did you fear?" Giancarlo was not so much upset, because he was wandering how to transform into patents something of what we had seen. Our friends were usually very amused with this peculiar habit of Giancarlo; he has been one of my closest friends.

Any way, we had got out at night, and went looking for the FIAT 600 that had been left there, and we did not find it. "Look, we have been robbed" I said, and our friends told us: "Don't worry, your car has been sent back to Giulio, who was needing it, and you are going to use our own car [13]." It was really a problem, because I had a driving license, but I have never liked to drive; I like motor-cycles, but I have never been used to drive a car; so it has been Giancarlo to drive our friends' car down to Pescara. When we arrived home, we remained there several hours discussing about our experience; the I realized that I was very hungry, and said to Giancarlo "Please, let's go somewhere to eat something." Giancarlo told me that he was all right, because, when starting from Pescara, he had taken a mortadella sandwich with him. "You have eaten a sandwich inside the base?" "Sure, was I to die of starvation?" This kind of guy was Giancarlo.

So we went to have lunch in a restaurant near to the sea shore, at Dino's. After lunch we went to the sea, and took a very long walk, first up to the harbour, then in the opposite direction, towards the North, up to Santa Filomena [14], then back again. At the end, it was night again. Hopefully it was a Saturday, so the next day Giancarlo had not to go to work.

Next day we met, together with Giulio, who was very upset in learning that we had been inside the base, without him. At first, he did not even believe us, thinking that it was a joke.

During 1957 we continued to meet our friends. In that year, Giulio had to move to southern America because of his job. Giancarlo and I were travelling frequently to Ascoli Piceno, but also to Como and to the Cadore, where they had opened some new bases. By pure chance, the owner of the timber warehouse in front of me was having commercial relationships with that region, because he was buying there his wood.

One night we had to go to Ascoli Piceno; as I was not driving, I asked a relative of mine to carry us there, and to wait some hours, to take then us back. That night we were admitted again into the base. When we were inside, our friends told us that we were to remain there until next day. "That's impossible; I have a rendezvous with a friend within a couple of hours." "Don't worry. We will fix the problem."

When a few days later I saw again that relative of mine, he told me that he had been waiting for us in the Meletti coffee-house in Ascoli Piceno, and that, at a certain moment, he saw a waiter who was going around announcing a phone call pending for him; he went to the telephone, and heard me (better to say, he heard my voice) telling him that there was no longer use for him to wait, because I would get back by myself the next day. So, he returned home alone. What a trick!

Next day, when we got out, my mind was in a strange attitude, thinking back to what had happened, but without stopping on single memories, as if refusing to admit as a reality what it was remembering. May be a psychologist could be able to understand this situation. We had been instructed to take a public bus to re-enter to Pescara, and had been told where to take it, how long the trip would have been, the cost of the ticket, and so on: our friends were aware of the transport facilities better than us! We arrived at half past eleven; Giancarlo at that time had no job, and was alone, because his family had gone to pay a visit to some relatives, so I went with him to keep him company; in the evening we went back to my place, had dinner, and remained talking.

About two a.m. we were sitting in the balcony, and a red cat joined us; it was an old acquaintance of ours, we had named it Miciolone (Author: something like "good, big cat"); it used to present itself in the strangest hours, very often skinned, with evident signs of fights against other cats; I remember that once it was so badly reduced that we had to take it to a chemist, and have it disinfected and bandaged, at the astonishment of the chemist, who was not used to take care of pets in the full of the night! Another time, I went out in the morning, leaving Miciolone in a nearby field; Franco, my nephew, who was living with me those days, told me that he had been watching the whole morning Miciolone, head to head with another cat; they were pushing each other, as if they were two bulls, and went on until after noon, when a lorry entered the field, and they were forced to flee away. It was a tremendous beast, but at the same time a very affectionate friend. When Dik grew older, the dog was standing the cat, but Miciolone had not to get too near to Dik, to avoid being barked at.

From time to time we were allowed to take some picture, either of our friends, or of their ships.

Dimpietro once told us "Look, now the CTR's are doing something, they are trying to hit us. Be very careful, and do exactly what you are told to do. The worst damage would be if they take catch of your mind. They are able to do so, it's quite easy to them; but if the mind reacts, opposes, nothing happens; if the mind doesn't react, it's done." "Is that all?" we thought; it looked too easy. I would rather have applied to the mental defenses I had become used to, thanks to my long practice of Yoga techniques, but he told me it was not necessary, it was sufficient to exert one's will, and to be conscious of it [15]. Giancarlo was always over-reacting in a striking way, and in this case he did as usual. We started noticing that, every time we went outdoors, two guys would be shadowing us, one walking before us, the other behind, thirty, forty meters away. It was always the same two men. "Look – our friends told us – these two men are almost mechanical; they are still men, but have been left with very little of their minds. They are under the power of electronic devices that control them, therefore do no try to harm them, because they are not guilty of what they do. Any way have no fear of them, because they are harmless; their job consists only in sending back their sensations, their feelings, so that their controllers are able to monitor the environment where they are. If you want do deceive them, show that you are engaged in a meaningless activity; they will try to get near, but they will not be able to understand anything; in the meanwhile, it will be easier for us to manage the situation". We did so; at once, we started reminding each other that we had to go to the sea shore, to look for certain metals our friends had left there for us; when we arrived to the place, the two guys were already there, waiting for us! We were pretending to haven't noticed them, and to be engaged in some mysterious activity, after what they disappeared, walking in different directions.

In those days, I still had my motor-bike, so I jumped on it, and went to the pursuit of one of these men, shadowing him from a distance; when he turned around a corner, I gave a short burst of gas, so I got there within a split second, but he was no longer to be found anywhere. It was very late at night, and the road had only closed shops on each side, so that the man had no place where to escape; nevertheless, he was gone!

In the country around Montesilvano, one night an hydra appeared (Author: "Hydra" is the name they were giving to such entities), a gigantic being, that was howling loud; it was very tall, like a building, and was shouting; there was a acid stench all around, and it was difficult for us to breathe; we had a long copper chain with us, that we had been told in advance to build. "Grasp the chain!" they told us, and in that moment we heard a crash, we felt an indescribable stench, and we heard a yell, you can't imagine how strong it was, may be the shout of a stricken dinosaur. Any way, there was something human in it [16]. And the hydra disappeared. We were astonished, unable to speak, considering what had just gone on. Raffaella, my wife, had remained at home, not far from there, sleeping with the young son of Carla, one of the ladies. She remembers that in the very moment the hydra disappeared, she felt our house trembling, as because of an earthquake. After some seconds, everybody started to talk: "Have you seen it?" "What was it?". I told them "Not only we have seen it, we have also heard it; it is strange that nobody in the surroundings has even realized what happened". We could still see something like a cloud that was being absorbed slowly, very slowly, by the ground, like a kind of fog; and there was still a smell somewhere between ammonia and muriatic acid; and a foam was boiling over the ground. "Please allow us some ten minutes to finish our job; do not walk over that foam, because it could burn your feet". We remained silent, there were seven of us looking at that show. Strangely enough, the place was desert, there was nobody, but us. And yet the shoot had been extremely strong, like a hooter, and the being had been very tall. Yet, apparently, nobody had been awaken. "Had we not be able to kill this being – they told us – within some ten days it was to destroy every kind of life all over your Earth" [17]. I had been stricken by that experience, if I were able to paint, I'd like to paint what I remember. It was night, there was no moon, and we were in the middle of the country, West of Montesilvano; now, in that area, many houses have been built, but at that time there was none, just bare ground; to get there we had been walking among earth clods. All of a sudden we saw this entity swelling itself from the ground; it looked like the tar used in roads paving, a smooth and shining tar that was moving, winding as fishes do, and inflating itself. And it became as tall as a building, yet nobody appeared to have noticed it, to my amazement. "How can it be?" I asked myself, and was answered "We avoided that the sound waves could propagate around, so that nobody has heard anything. Otherwise, the people from the village would have gathered here, all of them". But we have heard it, we were some 50 meters away.

Often they told us that they were needing food and vitamins, in great quantity. They told us to rent a truck and a trailer, to fill it with these goods, and to take it to Pineto, "so that we may unload it". I said it would have been a problem: how could they unload a truck, in front of its driver and eventual by-standers? "It's up to you to find a way" I was answered. We had always to find tricks to welcome their requests. I had to organize several transports like this first one, changing every time trucks and drivers. Once I choose a man from Foggia; I had bought two tons of fruits, because they had asked fruits (other times fish, never meat, because they would never kill or eat animals). Usually, but not always, they would take care of the payment, sending us the necessary money. I had asked how could they find Italian money, and was told that they had a special device, able to collect all the lost coins and notes, everywhere, all over Europe!

Another time, they sent us platinum ingots! That stupid man named Paolo [18] (who was later to leave the group) said, in that circumstance, that it was a kind of devilish situation, you see how twisted the human mind may be. It doesn't see what it doesn't want to believe. That time, we were in the garden of my "Green" villa, the one you know well, and the ingots had actually fallen from the open sky, so that we all were engaged in the task of collecting them from the ground. On the other side of the street (Europa avenue, in Montesilvano), a man was standing at his balcony, and I was wondering whether he could understand what was going on. We collected 10 boxes of ingots, about 150 kg of platinum altogether, you know, this metal is very heavy. Now we had the problem of selling them. In Milano there was our friend Emilio, who was working in the field of jewellery-making (his grand father had been a famous jeweller in Milano). He said that he knew a wholesaler of precious metals, and so we had been able to sell the ingots, strangely enough without any serious problem, without even being questioned about where they were coming from.

So we had been able to buy large quantity of different kinds of fruits, and charged two trucks with them. The lorries stopped in a pine-wood at the outskirts of Pineto, and then there was the problem of unloading them; of course, our friends were to do the job, via tele-transporting the fruits away, but how would the two drivers react at seeing their trucks all of a sudden deprived of their load? So I invited them to have a drink in a near-by coffee house, telling them that in the meantime some workers were to come and unload the trucks. We walked, very slowly, to the coffee house, and I was trying to waste as much time as possible. When we got back, the first lorry had already been unloaded, while the process was still going on the second one, a truck and a trailer: when we were some 30 meters away, all of a sudden we saw it jerking, because of its sudden reduction in weight, but hopefully I was able to persuade the drivers that it had been just an optical illusion! Then, when we arrived to the trucks, and found them totally empty, the two drivers praised the skill of the mysterious workers who had done so good a job in so short a time!

Another time we were again asked to send them great quantities of fruit: 5 tons of fresh fruit and 5 tons of dried figs; these latter I got from a wholesaler from Puglia (Author: Southern Italy), while I had to look around to find where could I buy so much fresh fruit; and again I found a person from Puglia, a Mr. Ronzoni; unfortunately he had not so large a truck, so I had also to find a lorry for the transportation. I paid in advance half of the bill, and told the driver to take the fruit to Pineto, on the start of the road going to Mutignano (Author: An even smaller village), where we would have been waiting for them.

While Giancarlo, I and a third person were waiting, we got a message from our friends: they were pointing to a specific place on the ground, near to our position, and stating that there they were to send us the money to complete the payment. Then Sigir said "Now", and on the instant a box appeared on the ground [19]. I have been present at such performances many times, but I never got used to them, to see an object appearing all of a sudden, coming from nowhere. The trucks were to arrive shortly later, and Giancarlo was very puzzled, because there were just the three of us, and no worker to unload the trucks; actually our friends had told us not to worry, because this was to be their job, but how would the drivers react while looking at such an unusual event?

When the trucks arrived, their drivers were a bit surprised to find no workers waiting for them; I told them than shortly five smaller vans were to come, with the necessary workers, and I proposed them to go, in the meantime, to a restaurant not so far away, to have some coffee; they accepted, so we left, leaving Giancarlo to take care of the operations; I told him to reach us when everything was over.

We sat at a table, and I was disappointed because from that place I was not able to see what was going on, there were some trees in between. I offered the drivers a small meal; from one side, they were badly needing a square meal, but at the same time they were eager to leave, because their next stop was in the Northern Italy, a rather long trip; hopefully, one of them started talking, telling me of some experiences he had had during the war, in the Pineto area, and I pressed him to continue his tale, asking for more and more details; we spent more than half an hour listening to the driver.

I told them that probably it was time to get back, although I wasn't sure at all about what had taken place in the meantime; they were a bit reluctant, saying that too short a time had elapsed, and that unloading was a long process. "Don't worry, the workers I have enlisted are a clever gang, I am sure the job is over". Indeed I was sure of nothing, but I was feeling urged to get back.

When we arrived to the trucks, we were met by a rather furious Giancarlo, who had got tired to wait for our return: "The job has been finished long ago!" I had told him to reach us after the unloading, but in all evidence he had forgotten, and got nervous while waiting for us; any way we found very clean surroundings: not a trace on the ground, even the inside of the trucks were looking as if polished. Only a vague smell of bergamots, citrons and lemons could be felt in the air. The drivers were astonished, they had never seen such a perfect job by a team of workers, and were praising how swift and clean they had been; of course, I too was praising these phantom workers!

On the average, I had to arrange such a play a couple of times each month, always with the same problems, the invisible unloading, the necessity to take the drivers away from the place, and so on; hopefully, everything always worked.

I was starting to get reputed as a kind of an queer wizard, because of my unusual behaviour, and was worrying for that, but my friends told me not to care too much. But also the persons nearest to me at times were showing strange attitudes towards me. Think for instance to Alberto, that drunkard (Author: This gentleman, Alberto Perego, had been an Italian diplomat all around the world); usually he would stand spirits for a while, but all of a sudden he would get totally drunk. Imagine that one night we were in the green room of my house in Pescara; there were Giancarlo, Giulio, somebody else, and Alberto; he had been my guest since some days before, but that night he was a bit over-excited; he started stating that he wanted to ask our friends several questions, and wondered how they were to answer them, because he would not accept a telepathic communication. A small sheet of paper felt from the ceiling, on it there was written "You have just to think at your questions, and we will answer in this way". Alberto thought that it was a trick from my part, and I was going to get furious for that suspect. He was a bit drunk, and told me to stop playing tricks; we almost came to blows, then he got quiet again. Then he went on "Let them amuse themselves", and started thinking of a lot of different questions. Within less than half an hour, the table was covered by hundreds and hundreds of sheets of paper, of different kinds, apparently torn away in a great hurry from larger sheets, and on them there were written the answers, in red ink, with an hasty handwriting. Alberto was astonished, and we too.

But in general, his attitude towards us was twofold: at times he was depicting me as a kind of a genius, other times as a cheat. He was a strange guy: do you remember that time when he was driving, totally drunk, and while turning. one of the wheels of his car went lost? Any way, he was not enjoying an happy life, had divorced from his wife, and was living alone in his house in Ruggero Fauro street, in Rome, with his daughter coming at times to pay him a visit.

In 1958, our friends had told me that there was also a base of theirs in Australia, a very little one, with very few persons inside, also four terrestrials with them; among these latter, there was a Henry Ford, of course not the famous Henry Ford; this gentleman was a German, to be exact his mother was German, his father was American. Any way, our friends were rather reticent on this subject; once Giancarlo tried to be insistent, and was answered "That's none of your business!"

Things were going on as usual; there was a villa on the sea shore, its name was Villa De Riseis, and beneath it there was a little base, and Sigir used to get out from there, at night, and we went strolling along the beach, together with Giancarlo; Sigir was a little more than 2 meters high, so that he could be mistaken for a terrestrial, for a basket player; and he was looking intensely at the sea, especially if there was the moon in the sky; there was a definite romanticism in that, not a romanticism towards women, but to the nature itself; he was saying that the nature was an epitome of the beauty of the Father, because he was very religious, but in a simple and natural way, without useless postures.

At the end of the 50's, Giulio had left Italy to work in Southern America; so only the two of us were remaining, Giancarlo and I; there were occasionally other people, whose name I do not even remember. Once two men came to us, a physician from Sulmona, and a lawyer from Lanciano; they saw a lot of things, and were as lucky as to enter the base under Pineto, and to stay there for one hour and a half. When they got out, they were unable to utter a word; "Now we cannot say that it is not true, but if we say that it's true, nobody is going to believe us. Now we can't get back." It was late autumn, they had entered late in the afternoon, and got out when night was coming. One was saying: "I told you that it was true", and the other "I told you that, were it true, that was going to ruin us, because now how can we live? We have a family, we have a job." Of course, these were egocentric discussions, instead of thinking of what they had learnt, what they had seen. Late at night they went away on their sport car. They had left us their telephone numbers, but insisted that we were not to speak of such topics by phone. "Usually it is my wife to answer the phone" said the lawyer; the other said "How to speak of such things by phone? By the way, my wife is never at home, because she owns a shop." They had promised to get back the next week, but they did not.

A few days after the miss, Giancarlo asked me what they were doing, and I decided to call them by phone. First, we called the physician, a woman answered in a shrill voice: "Are you by chance one of those who have hypnotized my husband?" "I have hypnotized nobody." "So, what do you want from my husband?" "Madam, the tone of your voice is out of place; please let me speak with your husband." "No, you must leave my husband in peace." and she shut off. We tried to call the other one, but nobody was answering. So, we decided to go to Lanciano to meet the lawyer. We found his office, but his secretary told us he was not in. "Fine, we are going to get back within half an hour." "No, he will not be back in so short a time." Just near to his office there was a coffee-shop; we went there, looking at the street through its windows; after a while, Giancarlo went back to the office, pretending to be a customer, and was welcomed by the secretary, who told him to wait just a few minutes; so Giancarlo, through a window, waved at me, I came back, and he opened me the door. When the lawyer saw the both of us, he told us that a real catastrophe had happened to them; they had decided to tell the story first to one wife, then to the other, but both times being together; they had started with the lawyer's wife, and she had reacted in a furious way; she assaulted them, then, the day after, went to the police, pretending that her husband had been hypnotized by persons who had him seeing untrue things, for some unknown reason, but probably not a honest one. "You only look as two men – I told them – but you are not; you are just two entirely dominated persons; your wife is hysteric, a greedy woman. Farewell." We never met again.

As time went by, I told Giancarlo that I was tired of such situations: "I am looked at as an hypnotist, you as my assistant; it is like a theatre piece, that could have been written by De Filippo (Author: A famous Italian comedy-writer). Please, do not accept anybody else into our group."

Later on, we had to go again to Ascoli, because we had to meet some newcomers; from eleven o'clock at night we remained waiting outside Rocca Pia; it was winter, it was raining, and we could not find a way to stay comfortable; for a while, we took turns, one of us sitting inside the car, the other waiting near to the castle; at half past three, we saw a brilliant light, but it was a strange light: it was a solid light cylinder, a couple of meters wide, and may be three meters high; strangely enough, this light was not illuminating around, around it all was as dark as before. When later on I asked our friends about this peculiar light, they were joking at me: "Haven't you yet understood?"

All of a sudden, a person rose inside this cylinder, and got out; then another one did the same; at the end, four of them had got out. When such things were occurring, I was feeling uncomfortable, and Giancarlo was doing the same; only, he was recovering immediately; I have always envied him for his ability; my peculiar attitude was different: I was always studying myself, studying what was going on, to get sure that I was not dreaming. You know, in front of such things never seen before, almost unbelievable things, you tend to switch between reality and daydream, at times you believe you are just dreaming, and you have to convince yourself that it is true. Any way, we cheered these four persons, and they told us that they were to leave after a couple of weeks, and would get back one year later "when you'll live in Milano". Actually, there was a vague plot for me to move to Milano, but at that moment nothing was for certain. We remained a long while discussing; in particular, they wanted to get sure that we had really understood who they were, that there was no terrestrial origin in them. Actually I have had such an idea many times, during the first years, but then I was forced to convince myself that it was not possible. We remained there until dawn; when the sun rose, we said good bye each other; Giancarlo said that he was willing to get into the base with them, but they said

"No, it is up to us to decide when you may get down; then you may stay as long as you like; but now it would be dangerous to you to get down, so forget about it."

It was no longer raining; there were stars still visible in the morning sky. Giancarlo told me that he had a sandwich in the car, and he went to eat it. I got a coffee at the Meletti bar, that had just opened, then we returned home.

Giulio was not with us in those days; he was traveling through Italy because of his job; in that time, he was negotiating with Dalmine (Author: One of the major Italian steel companies) because he was needing a great quantity of steel pipes. Before he left, we met, and we talked at length about what we had experienced. He had seen a lot of things: objects that were "taken" by our friends (that is, disintegrated in order to be transferred inside their base), or things that were brought (the inverse process). He was not willing to move, but I told him that it was necessary, if his company would pay him enough for that. So he left. The two of us were sad, even Giancarlo, although he had often quarreled with Giulio. Moreover, we were left with no car, because up to then we had almost always made use of his car.

Things went on. The contact was going on, it was a beauty; very often they were saying us: "You see, my friends, if you do not know the alphabet, how can I teach you the language? I may do my best, but there are things that I cannot explain you, simply because you are too ignorant. Up to a certain level, it may be done, but beyond that it is simply impossible". "That is not so important – I was answering them – you presence here, these contacts, are more than enough for me."

The Milano schedule was approaching; I had asked my sister, who already had a job in that town, to find me an apartment, and she had succeeded. So my wife and I started to collect our things, to buy some furniture, to start our new life in Milano. Our new apartment was in the San Siro area, because our friends had specifically stated so. At first, we had problems with our two sons, because they didn't like the new place, and were always arguing with their new acquaintances.

So, by now, the Milano stories were starting; in the San Siro area, near to the ring-road, there was an entrance to an underground base; it was on the top of a very low hill, and I was often going there to meet our friends. Giancarlo had got a job in Rome, and usually would come to Milano during week ends; therefore, most of the times, I was alone in my contacts with our friends. Any way, on a Sunday, we were supposed to get together to Como (a town some 50 kilometers North of Milano, on the shores of the lake that has this same name); I phoned one of my relatives, who was living in that area, and asked him to drive us to our destination; to my surprise, he answered that the place where we had to go was very near to his house! In that area, on the lake's shore, there is an ancient villa by Palladio (Author: He is a famous Italian architect of the XVI century), the house is almost desert, may be only its keeper lives there. In front of this villa, some steps go down under the level of the lake's waters.

As in the famous De Mille movie about the Bible, the waters spread away, and a couple of our friends came out. We remained talking for more than one hour; by the way, they listed to us the items we were supposed to supply them; and at the end of the meeting, they gave me a small object, some 20 cm long and may be 15 wide, and told me to take it home, telling me that thanks to this device I would be able to get in touch with them whenever I'd liked to. Its surface was flat and dull, but, approaching an hand, four buttons were to appear; one was to tell our friends that I wanted to communicate with them, another to establish the link, a third one was some kind of an alarm, and now I do not remember the use of the fourth button; taking the hand away, they would disappear.

Many times I met Sigir, Meredir, who was everywhere, and was always escorting Sigir; also Dimpietro was there, in the northern area of the base, so that to find him I had to use a different entrance; he was living by himself; his room was very wide, but not so high, so that the head of Dimpietro was at a scarce half meter from the ceiling; there was a large table, full of different lights, in total silence; Dimpietro told me that it was a device that allowed him to manage all the facilities, and all the people, inside the base. He also told me that they were not afraid of the great amount of waters over their heads; on the contrary, they were worried by the fact that very often on the shore happenings were taking place, with a lot of people attending, so that it was often difficult to them to get out and back, least could they be noticed. Moreover, in the Como lake there is a seaplane station, with planes flying around at all times.

The winter passed, now we were in 1961. On a Sunday Giancarlo got back from Rome, and told me that he had met a young man from Turin, and that I could be able to see him the next Monday on the TV, because he was conducting a transmission for youngsters about model aircrafts; he was a very good and very intelligent guy, Giancarlo told me, he was also a journalist, writing in a pleasant manner, usually about technologies, aviation, modelling, and the like. By the way, he too was named Bruno.

I looked at the broadcast, I liked this man, and told Giancarlo to invite him to my home; you see, when things have to go in a certain way, before or later they will arrange to do so. I met this man, we had lunch together, and I started telling him what was going on, to his extreme astonishment; he was then accepted into our group, and started sharing our experiences; when he was in Pescara or in Milano he usually slept in my home.

He made recordings of some of the speeches of our friends, and had them listened to here and there, and started writing articles on this subject. Our friends told him that he had better to stop, or at least to pay great attention, because most of the people do not like to accept this reality, and so he was going to forfeit his prestige. Actually, he was pressed by many publishing firms, Mondadori and Rizzoli (Author: Two of the major Italian publishers) among them, whose managers has seen our pictures and had understood that the matter was an important one; the director of the newspaper "Il Tempo" asked Bruno to write one article a day, during a whole year, on the subject; at first Bruno tried to refuse "How can I? This phenomenon does not depend on me", but the director answered "It's up to you. Just find the way!" I did not like this demand, so I forbade Bruno to write anything about our experiences, because I didn't want they to be exposed to the meanness of people. Any way, Bruno ran his daily rubric writing of general UFOlogy, without quoting most of our cases.

Now we are at the beginning of 1962. We were always making long walks, often along the sea shore, and I found out that my friends were looking at me as if I were a saint, a prophet, and I didn't like that at all. Also my dog, walking at my side, usually was forcing its head up to look at me, and listen to my talks! It is true that while I was telling what I had learnt from our friends, I often got so engaged in the process, that I was not even realizing what was happening around me; once, I remember, while walking and talking along the sea shore, I entered the waters, and continued walking, without noticing that my legs were wet; Giancarlo then stopped me, crying "Do you want to do like Jesus Christ, and walk over the waters?"

Giulio was no longer with us, and we had a large interaction with him via mail, telling him what was going on; at a moment, he told us that probably he was to get back, because his business there was going to settle, that he would have liked to get back any way, in order to get again in touch with our friends.

Once we were told that we had to meet Siderius, one of the W56's about whom they had talked to us a lot; he was something in between a scientist and a philosopher; we had never met him before, so that was an important occasion. Three or four days in advance, Sigir, Itaho, Marius and all the others, were speaking about this occurrence. One night, all of us were eating pizza, together with Dimpietro, who, as usual, was joking and mocking at everybody; great men are enlightened with the gift of a true witticism; we were discussing about what was to be done. It looked that Siderius was to arrive at half past two in the morning, and we were to meet him in a secluded place along the banks of the Pescara river. Giancarlo was very excited. All of us, both the W56's and us, tried our best to find a way among the vegetation and the walls that are present along the banks; you may imagine that doing such a thing at night was not an easy task. Any way, we succeeded in getting to the desired spot, just on schedule, and we saw a white sphere, some 20 cm in diameter, suspended in the dark; Giancarlo at first thought it was the lamp of some fisherman, but, as we got near, the sphere disappeared, and we saw Siderius, and two other persons, sitting in the dark waiting for us. At first, we went back to the sea shore, where we had parked our car, then Siderius said that we had to go at once to Ascoli Piceno. It was not an easy task, because there were nine of us, three of which were extremely tall, and it was not so easy to put everybody inside our car! Hopefully, they had also their cars, so that, in a way or another, we were able to go to Ascoli. When we got to Rocca Pia, we entered the base (to me, it was the fourth time), we went down, and entered a new room, that I had not seen before, and got seated around a table, on some kind of cubic chairs, without a back.

We were offered some kind of sweets, that Giancarlo refused because, he said, they were dangerous to his teeth; then Dimpietro ordered pizzas for everybody! To our astonishment, he told us that he had read many culinary books from almost every European and African culture, so he had taught his fellows to cook whatever he liked. While eating we have been speaking at length; Siderius spoke a perfect Italian; when we speak our language, usually we force it in a way or another, we add a bit of vernacular; on the contrary, his language was really perfect. He spoke in a calm and regular way, as if he had learnt by heart what he was to tell us; he spoke about the universe, about the invisible worlds, that are finer than the visible one, about death, saying that one must get ready to death, instead than life, because life goes by, while death arrives to everybody. He spoke about the CTR's, and, from time to time, he was addressing his fellows in their own language, but he always asked us the permission in advance. At a moment it looked as if he and Dimpietro had something they were not agreeing upon. I told my friends "What, they too do quarrel?" But there was no real quarrel, just a kind of tense interchange. Giancarlo said "No, that's just Dimpietro who has got drunk, and now utters rubbish!" Of course that was not the case, I have never seen Dimpietro drunk. Any way, we had not been able to understand what was the matter. The situation settled itself quickly, and after a while we were told that we had to get back home. They greeted each other putting their arms around the other's neck, then all of them did the same with us: "Do you understand how much do we love you?" Dimpietro said.

It was a quarter past four. We had left our car in Pescara, and so they told us that we were to be taken there in a car of theirs, driven by one of them I had never seen before. "Do not try to speak to your driver – Dimpietro told us – because he will be deaf, dumb and blind; he sees, hears, and speaks quite well, but he has been instructed to behave this way." So the two of us, plus our driver, got out, entered a car parked not so far, and started our way back. Actually the driver was very strange: he was looking directly in front of him, not trying to interact with us in any way; he was driving calmly, not too quick, in a rather pleasant way. Giancarlo and I were asking ourselves whether he was an actual person, or a biological robot; at a moment Giancarlo told me that he was going to prick him with a needle to see what would happen, but I prevented him from doing so. Any way, we arrived to my home; before getting out I said him good-bye, and he just nodded.

When I went to stay in Milano, Bruno, the journalist, moved there his activity, in order to stay near to me, actually living in my home. Also other people were getting in touch with us, thanks to the papers written by Bruno. But I always reserved me the right to chose whether or not to accept anybody. Among this people there was Gaspare, his wife, his mother-in-law, who at that time was already over eighty, and an old gentleman, Nino, who was the second husband of his mother, and the son of Gaspare. Gaspare was a painter, a short guy, very witty; at times, he was also giving tennis lessons; his wife was a rather tall woman. Later on Emilio joined us; we had nicknamed him "Seghè" (Author: That's an almost untranslatable expression in Milanese vernacular, meaning, roughly, "What's that?"); other people arrived and were sent away, because I felt that they would be of no use to us, so I excluded them, and obviously, they started venting their spleen at us after that.

One afternoon, a couple of flying saucers appeared over Milano. Our friends had told me in advance of this flight, so I had asked the permission to take some pictures. We went to a terrace on the top of a building in Giulio Cesare square, and I was able to snap my pictures, that were also published on some newspapers, because especially one of them was very clear, with Milano skyscrapers in the background, and the disks directly over them. Many of us were looking through binoculars, although one of them, when the sightings was over, pretended that the saucers were simply some high flying birds. For many of them it was the first time they had sighted a flying saucers, so they were a bit upset.

A couple of years went away. Gaspare and his family at first were living in Giulio Cesare square; then he bought a new house on the shore of Como lake. During summer, they used to come to Montesilvano to spent the holidays with me. In summer we were often going to Pineto or to Ascoli to get in touch with our friends, or were receiving messages, either telepathically, or via radio sets. Carla, the wife of Seghè was affected by an almost lethal blood disease, hemophilia or something like that; she was very pale; her physician was not able to solve this problem, so that at a moment Sigir decided to use their capabilities; he told her: "My daughter, we are not able to act this way on everybody, nor are we sure to meet always a success; nevertheless, I have studied your problem, and I am sure that I am able to cure you in two, three days." The lady started to cry, you know, when all of a sudden an help comes for an apparently unrecoverable situation, you get really upset. She was asking me "Is it true that he is going to cure me?" "I have seen that so many times he has done so, therefore don't worry." Very often, ill people were cured without even having being aware, so they looked to be recovering by a miracle. Of course, our friends did not want that this capability were widely known, and so they usually were acting on a secret base, elsewhere I was to expect a queue of ill people in front of my apartment; actually, at a moment, I thought to make a business on that, but I rejected this idea almost immediately, and my wife, even now, reproaches me for that: we have always been in a keen want for money, and I could have earned some money from this business, had I entered it; unfortunately, I didn't. Any way, Seghè's wife recovered, and I still have an audio tape from which you may hear her crying, and thanking Sigir for what he had one. Every time there was a contact with Sigir, and Carla was attending, she always was bursting into tears, so that once I told her to stop, or else she was going to flood my apartment!

Later on Emilio got a problem with Carla; one day Sigir called her, and scolded her harshly; from that moment, she got very angry with our friends. So the peace within our group was starting to disappear. Another time, a couple of brothers, both working in a bank, and both very short, stole from my desk an audio tape, and a film; I was very worried, but Sigir told me to be quiet, because the CTR's had tampered with the film by far away, and now it was useless: in its original form, it was showing a couple of flying saucers moving in the sky, and passing behind some chimneys; after the tampering, the disks were appearing in front of them, between the chimneys and the operator! After all, we got some good from this experience; before it, very often we were shadowed by, may be, government people, G-men, probably looking for what we knew; after that, probably people started to believe that we were a lot of cheats, and the pressure upon us started to lighten. We had made a name for ourselves!

In 1964, I was still living in Milano, but had already built my "Villino verde" in Montesilvano. A lot of persons have been hosted in there, among them Walter, a Swiss gentleman, Verena, his wife, and their sons, Marino and Maya. I convinced Walter to give one of his cars to Giancarlo; this car was a Renault, a convertible car. This car lasted no longer than a couple of months. One day, Giancarlo had to go to Roma, because at last he had been able to buy an apartment, and he wanted to furnish it; before leaving, he had taken away from the car everything that was not strictly necessary, transforming it into a kind of a small van. He had just married, and went to Roma to take all the presents he and his wife had received, and also some piece of furniture, overloading this way that small car. It was like when you look at a comic, he, his wife, and all this stuff over that Renault. Just to add another problem, while getting back they ran out of fuel, and it was a problem because it was a Sunday (Author: In Italy, on holidays, Sundays, and the like, typically most gasoline stations are out of service), and he had to make a long walk, taking an empty can with him, looking for a service station. Hopefully, he was able to find one, just a couple of kilometers away, where he got his can filled with gasoline. Any way, the car did not survive a long time after that disastrous trip: first, a spring broke, and was repaired; then, it ran out of oil, because the meter was not working, and Giancarlo simply forgot to test the amount of oil inside the motor! Therefore he was forced to dismiss it, after having taken away all that he thought could be useful.

Once I asked a friend of mine, father Filippo, a friar who was living in Cepagatti (a small village some 20 kilometers west of Pescara) to host Giancarlo for a few months in his small monastery, so that he could attend in peace to the development of his technical projects; from time to time I was presenting father Filippo with small offerings, in change of his hospitality. Later on father Filippo got out from his order, to become a priest, because he was in heavy economical problems (Author: In Italy friars live on charity, while priests receive a regular wage by our government), and I was so offended that I didn't want to meet him again. Any way, in this monastery, Giancarlo had been working to the design of a kind of weapon, with the technology of our friends, and one day I was invited, together with some friends, to be present at the first experiment of this new device. Father Filippo was worrying about possible damages, but Giancarlo told us that everything was under control. After a long preparation, Giancarlo operated his weapon, aiming it to a nearby tree. I must confess that I was rather skeptical; after the shot nothing happened, and we went to have a closer look at the gun. Giancarlo noticed that a gold needle, that was vital to his device, and that I had bought for him, had become like ash. "This needle has not endured the stress!" he was saying; we got back inside the monastery, but after a while were called back by the friars: "The tree has caught fire!" Evidently in a way or another the weapon had done its work.

Then Giulio returned to Italy. He went immediately to Giancarlo's, and they came together to Milano to meet me. My small apartment was full of people: my family and Bruno, the journalist, any way we were able to accommodate these two more people. We were told that we had to go to Pineto, because an operation was to take place, and it was necessary that as many persons as possible should attend. So we went to Pescara by car, there were 10 of us, and went to stay for the night at the hotel Dino, in Firenze street; early in the morning, we went out, one after the other, arising the perplexity of the night watchman: "Where all these people come from? Are they, may be, leaving the hotel without setting the bill?" "Don't worry – I told him – I am a friend of Dino's, and later on I'll get back to fix it."

Then we went to Pineto. It was winter, and there was a lot of snow, so that it was difficult to negotiate the hill by car. When we arrived on the top, we remained waiting; Bruno stated that he had a tape with a speech made by Dimpietro, directed to all of us; so he started walking up and down, in his camel overcoat, aiming at us his recorder and playing the tape. In the mean time we received a message from our friends, telling us to place ourselves in a certain way over the ground, and telling also that they were to take away the material we had carried for them, but this time not from under the ground, but from the sky, because, for some unknown reason, they were not able to make the operation directly from within their base. We remained there for over one hour and a half, it was very cold; all of a sudden, we felt the ground trembling, we saw a fierce light, emerging from the ground and speeding towards the sky; then the light came back, disappeared under the ground, and we heard a loud crack. After a while, small leaves started falling from the trees. "Now you may leave". "Is that all? " Bruno asked, and Giancarlo rebuked him harshly. So we went back, but many of us were confused, because they were non able to understand what had gone on. In the following days, back in Milano, these people started to loosen their relationship with us, they were making silly jokes via telephone and we never met again many of them.

Then Giulio started to suffer for a pain in his spine; our friends made a kind of remote diagnosis, then asked him to place an unexposed photographic film, in the dark, around his back. When the film was later developed, there was a sort of radiography impressed on it, and they told him that the problem was consisting in a displaced vertebra, and that an hernia was impending. "Any way, we'll fix it, there is no need for you to undergo an operation." At first, Giulio looked to have accepted the proposal, but a few day later he got hospitalized, and operated. When I heard about that, I was astonished, and the worse has been that, after having being discharged from the hospital, he was feeling even worse than before. Sigir told him that in this new conditions, they could be of little help, because surgeons had damaged some nervous terminations, and he was risking to remain paralyzed in one leg. They sent him a very thin copper plate, and told him to apply it over the place his spine was aching, and to keep it fixed there during a whole month.

Once, in Milano, many of us had gathered together. We had been told to get to a secluded spot, where three ships were to land. It was winter, and the morning was very foggy. All of a sudden, the smog looked to condense on its sides, leaving an opening in the center, and three flashes zapped through it, then everything returned normal. We walked in the direction of the lights, and in a short time we arrived near to the disks sitting on the ground. They were at the center of a small valley; we were prevented to get too near to them "We are not here to make a show, we have our own business, but if your friends like to have a look, they're welcomed; only, please, do not get too near to our ships. We may give you ten minutes at most, then you'll have to go away. And do not take any picture! Your cameras would get no result."

One of our group said "We are going to see if that's true", and started to snap photos. I got angry, because I had transmitted them the request of our friends, and told him to stop at once. "I am already done, tomorrow I'll give you the printouts." Of course, next day, there was nothing in the pictures, as our friends had told us [20]. "How can it be? May be it has been just a dream." At that, I drove him out!

Getting back to the previous day, our silly guys got satisfied: they had the need of seeing something, and our friend had complied. Any way, when the ten minutes elapsed, they transmitted me the urge of being let free to attend to their business, because the show was over. We remained there some minutes longer, because some in our group did not want to get away; we started to quarrel, and, all of a sudden, the three ships disappeared! I got the message "We are still here, only we had became invisible, so that your friends will get tired of staring at nothing." So I told my fellows "OK, they just went away, now we may do the same." "How, they were here one second ago!" "Well, if you do no like, you are welcomed to complain!" I had got really furious at those stupid men. Through such episodes, I was starting to realize that there are too many silly persons around, and that probably it would have been better not to admit other persons inside our group.

Later on, Giulio apparently had a quarrel with Bruno, and Giancarlo came to me, saying that strange things were taking place, that the two of them were looking to be changing their minds, were saying strange things, as if something unforeseeable had happened. I was not able to do anything, I didn't even know what to do; they would not listen at me, because they had really changed their minds, and would not accept anything different from what they were now believing. They came to me, and I got very angry during the ensuing discussion, at a moment I was going to break a bottle on their head! Giancarlo stopped me, they flew away, and since then we had very little contacts with them. Later, our friends told us that now they were under the yoke of our enemies, and that when one enters such a situation, it is almost impossible for him to get out from it.

Many years later, I went to Bruno's, and he was actually touched; he took me into his bedroom, and showed me three trench coats that in the old days had been charged in some way by our friends, and he was still keeping them with a great care. We started talking, and he was regretting that he was feeling that this story was not to reach any result, that it would go on this way endlessly, and he was not able to understand why we were still engaged into it. What was I to respond him? "We are not living an adventure, that has to end at a certain moment." Once again I got angry, I went away, and haven't met Bruno since then.

Paolo, although he was a good friend of Bruno's, remained with us; for a while he moved abroad, and the very night he got back to Italy, at 2 a.m. he came to my apartment to greet me. That night my family was not in, and beside me a friar was sleeping. He had been sent to Milano to attend a convention, and was staying at my home. Paolo was rather astonished; as he was tired, I made a coffee for him, and we remained talking up to the morning. Then he went to the station, to take a train home. He was employed at the FAO in Rome, had a degree in Economy, and we had met because he was working with Giancarlo's wife.

Then, also Paolo left, in a very stupid way, in 1970; during these years, he had been with me almost all the time, coming with me during my trips, continuing to share my experiences, and most often I was paying his expenses. He had been telling me that he was going into raptures, while he was listening to our friends, that no university professor, no text book was at their level: "They are telling us extraordinary things."

In 1970 Dik died; the same year I was introduced to father Domenico, who used to speak strictly in dialect: "Hey, boy, it's a long time I've been looking after you – he told me when we met – I have been looking for you, but have not been able to find you up to now. By pure chance, some days ago I've met the veterinary surgeon who usually comes here to Manoppello, and I asked him whether he was acquainted with a man so-and-so" Di Biagio answered father Domenico: "The only person I know, who is very religious, friend of so many friars, author of so many books, is Prof. Bruno Sammaciccia."

After father Domenico had been able to find me, he started frequenting my home, giving me many tasks, typically to help him in his historical researches on religious topics, and to give lectures, which I was really pleased to do. And his blue eyes were always smiling at me; I remember his long white beard, he was looking like Santa Claus, and indeed I was used to name him "Babbo Natale" (Italian for Santa Claus). My friendship with father Domenico lasted seven, eight years, then he died because of a car accident, in Turin.

Again in 1970 I was met by Sadi; his true name was Assad; his father was an university professor of Islamic theology; his parents had divorced when he was 7, he had studied in a Lebanese college (he had been born in Lebanon), then he came to Rome to attend university. When he was taken to my home, he was, may be, 23. He was a Judo champion; the day we met, he was strikingly limping, because he had hurt one of his knees during a contest. I started to explain him our situation, which was extremely strange, of course, but he was accepting everything without getting too much upset, without reserve, to the point that it was me to be a bit perplexed. "Have you understood what I've told you? It's not all made up, has nothing to do with politics, religions or the like, we are not a sect, it's just a very unusual, but immanent experience, an all-encompassing one." "Yes, I've understood."

He was, and is still now, very intelligent. "It's so strange, I feel that you are speaking about a great reality, and I would like to get a deeper insight into it." He was making use of a very simple psychological trick, staring at me to understand whether I was telling the truth, and of course I was aware of that. When he got convinced, he stopped that attitude. So he started to share our experiences, to meet our friends, to act according to their requests. Very often he and Paolo were together, but Paolo was starting to feel a bit nervous, uneasy in front of a world that was beyond his understanding. He told me that he had spoken to a priest, a professor at the Cattolica university in Rome. "You are nothing, in respect to these people – he told the priest – when they speak, they tell concepts that have never been heard of before. The best lecture by our best scholars is nothing, in front of what our friends tell us." "Care, because tricks may come from lower worlds" the priest answered. "But no – said Paolo – these are not captious speeches, they tend to elevate us, not to the other way."

Once Paolo had an experience that has shocked him; he didn't want to tell me in full what had happened, just that a man had appeared all of a sudden, uttering "Now you are in my power", but I know nothing more about that accident. Any way things were getting worse and worse with him, when I looked at him he usually was turning red. "Here there is something brewing" I told to myself. Then, all of a sudden, Paolo came to me, telling that someone had explained him what was going on, in a rather simple way: of course, it was a case of diabolical possession! "You are going to see me no longer, I can't participate in a diabolical affair." Sadi, who was living in Roma at that time, told me later that he had met Paolo, and had beaten him harshly, because he had said slanders against my wife and me. Sadi, who now teaches at university, was given various kinds of tasks by our friends, to collect the goods they were needing, or to sell precious items to get money for other operations; these have never been jewels, but unrefined materials; for instance, I had never seen a gold nugget, and when Sadi did show me one that he had been given, in order to sell it, I was not able to recognize what it was: I would have said that it was oxidized brass rather than gold, but when they had it melted, actually fine gold resulted.

When we were given lessons by our friends, Sadi was usually listening carefully, and was understanding almost everything, because he was very intelligent, but I noticed that his was a cold intelligence, like the surface of a frozen lake at night, under the moonlight. I was doing my best so that he could open his soul, start to make use of the superior parts of his being. He was usually practicing Yoga, so in a sense he would not find too many problems in doing so. Little by little, indeed, he actually started putting his rationalism aside, not forgetting it, but adding intuition to it.

Then, Sadi got his degree, *summa cum laude*. Often he makes me smile, because he is really very fat, he loves to eat, and he eats really more than you can even imagine. Once the two of us went to Losanne, in Switzerland, where I was to give a lecture, and in the morning we went to the hotel cafeteria to have breakfast; through the windows, I noticed a magnificent church just beside the hotel, so, as usual, I went out to have a better look. When I got back, I found that Sadi had almost emptied the self-service buffet table, to the astonishment of the waiters, who were staring at him! I was feeling a bit ashamed …

I really love Sadi, and he feels a kind of reverence towards me; in his books, he often writes about me as "Papà Bruno" (Author: Italian for "Bruno, my dad"). In "La via della realtà" (Author: "The Path towards Truth") he recounts covertly about the concepts he has learnt from our friends; actually, he had to write two more books on the subject, because I had told him that this first one was not so easy at all for unaware people to understand but, to his dismay, they did not meet with success. Very few copies had been sold, indeed I presented many monasteries with most of the remaining copies, and Sadi was very disappointed. "If you write things that most people are not able to understand – I told him – it is not you who has not met with success, but it's them who are not up enough."

Our friends had told me to build a large villa, on the top of a high hill near Montesilvano; I may say that the global design, and the actual building of this residence had been made under their guidance. It was really very large, on three floors, with many meeting rooms, large convention rooms, cubicles for individual studying; there was even a small astronomical observatory on the roof, and, obviously, within its foundations special arrangements had been designed, so that our friends could emerge from their base. I was to gather inside this house a large amount of people, and to introduce them to this new reality, having some of our friends with me, to assist me in these operations. To say the truth, they had told me that my new house was to become their operational center, and that they would give us some technical projects that we could sell, to get some money.

Among my guest in the villa there was the Swiss family: the husband, Walter, was an engineer, his wife Verena had a degree in foreign languages, but never did she teach; she was very good at driving their jeep over rough ground; they were living in Aurillio, a small village near to Lugano, in the Swiss mountains, and a jeep was required to go up and down those cliffs. Then there were their sons: Maya, their daughter, was a clever commercial artist, a very esteemed one; she was also a sea diver, but our friends had prevented her from going on, because, they told me, such a practice was going to speed up a latent disease; actually, when she was going very deep, she started hearing a music, some Beethoven, and when our friends heard about that, they stated that is was a bad omen. The girl started to have problems in walking, her sight too was affected; she got hospitalized, and was diagnosed a multiple sclerosis; I called Rita, a very good friend of mine, and a profound expert in biology (Author: I do not report here her family name, but am able to confirm that Rita is an authority in the field). She told me that there was nothing that could be done. "Within two, may be three years at most, your friend will get immobilized into a wheel-chair by a paralysis; nothing can be done to prevent that. At times, seldom, people in her conditions do recover, but our science is unable to understand why." Therefore I sought help from our friends, and was told that even to them it was impossible to have her recovered. "All we can do is to slow down the process. She will start to deteriorate within three years or so, then will get worse and worse during the next 13 years. We are not able to predict now what will happen at the end." Things went actually this way [21].

Then there was her brother, Martino, very fond of electronics; he stayed with us for a long time, attending the university in L'Aquila, but was not able to get his degree. Now he works in Lugano.

Verena and her daughter now retired into a boarding-house for elder people; when she calls me by telephone, she usually tells me that her body is in Switzerland, but her thoughts are here, with my wife and me.

This Swiss people were, at the start, Calvinists, then I converted them to Catholicism, and introduced them to many bishops, to father Domenico, that really loved them very much. They have lived with me for many years.

At a time our friends left for a short period. It was May; Martino, Giancarlo, Walter and I were strolling as usual along the sea shore, when we got a message: "We have to tell you something important; please proceed further North." We were roughly in front of Umberto avenue, so we proceeded; a couple of kilometer further, and we got another message: "Fine, now you may rest; in that place there are no unfriendly ears; for reasons we can't explain you, we are to leave for a period of seven, eight months; do not worry about the CTR's, because before leaving we'll do something to prevent them from hurting you. After this job will be over – they went on – there will be a change in our activity, and we'll start a new work to elevate not just you only, but the whole of mankind." They also gave us a set of instructions about what to do if we were needing some help. Actually they went out for seven months and a half. Every day Giancarlo was in unrest, saying that he was not able to stand this situation, and was asking me to do something, but of course there was nothing I could do. From my own part, I too was very sad, those seven months looked to never come to an end; to us, to get in touch with our friends was a vital need, something like the habit of smoking: think if, for whatever a reason, you should stop smoking for seven months (Author: Both Bruno and I are rather heavy smokers). Usually in the morning I was to receive a message from our friends, most times just a "Hi!" because most of times there was nothing impending, and now all of a sudden it was silence! After seven months and a half had elapsed, Giancarlo and I were sitting in my cabinet in Genova street, and we got a message from our friends, stating that they had got back! I am not able to describe what took place inside my flat!

In the first days of April, 1972, Giancarlo, Sadi, Gustav and I went again into the base under the Monti Sibillini, and we remained there for two or three days. When getting out, we all were surprised, because our clocks, and the outside world, were telling us that some days had elapsed, while to us it was looking as if we had been inside for no longer than one single day. Our friends then told me that inside their base the gravity was a 20% less than usual, therefore one could move easier, the heart beats with a less strain, and that was the cause of our mistake in the evaluation of time. They said that there was a technical goal in this lower gravity, but I stopped their explanations, because I was sure it was not in my capabilities to understand any technical detail.

Then we were admitted into other bases; they were all alike to each other: one was always struck at first by the all-pervasive light, then by that strange feeling when breathing, it was as you were feeling you were acquiring energy at every single breath, a bit as while you are practicing the Yoga "deep breathing", but the feeling was much more intense than that.

Then the catastrophe struck, the CTR's had got hold of the most of us, and probably the main cause has been a woman, the rotten apple because of which all the other apples in the basket get rotten. Our friends had been continuously goading us, saying "Dear friends, remember that unity is strength, if you fight against one another, then the adverse forces and our enemies are going to take advantage of that, your group will disintegrate, and we will be unable to do anything to prevent that. A group is like a living entity, and when it starts to break away, all the projects it is pursuing go astray. And then there is nothing we can do. When your defenses are weakened, our enemies may take grasp of your minds, change your memories and your will. The only way to prevent that consists in keeping together, with a good Uredda (Author: In their thoughts, Uredda is the general harmony, it is not just an abstract concept, but an actual living entity that operates on persons, on situations, on everything), you may keep your will, and protect yourselves."

I started to do whatever I could in order to keep our group compact, I was even imploring all of them to keep quiet, to remain firmly together. Every communications from our friends was always ending with the directive "Stay together. Had you to complain for anything, please tell Bruno, or tell us directly, because on the contrary, these negative feelings will end up prevailing unto yourselves, and unto our common history." But all that was to no avail. I was often calling everybody to my home, and was asking them if anything was wrong with them; they were always pretending that everything was OK, but our friends were then telling me that they were lying; and I too was aware of that, but could say nothing, least they were aware that our friends were controlling their sentiments.

So problems started to arise. "Look – I told them – if you start breaking Uredda we are going to face an enormous problem." A night I called Giancarlo: "Giancarlo, listen to me: did anybody complain on anything? Is there anybody who stirs up, who sows doubts?" Unfortunately, Giancarlo too was very puzzled.

Then, one after the other, they started to get away, each one with a different justification. One told me that he was feeling as if he were living a cloister situation, and he didn't have any vocation for that. There were many different tunes, but all of them were coming from a single orchestra. It was evident that nothing was working any longer;

also my business was going badly, a lot of complications every where, a real mess.

They were telling me: "You have the soul of a prophet, the mind of a mystic, but you are encompassed by situations that require the ability to manage other people, and you must take care of that." Was I really the supreme leader of the group, was I to impose my will with a whip? That's foolish, we may just joke on that. Probably I have been the spiritual leader of our group, but I have never imposed my will unto the others.

Our colleagues were not able to adhere to the teachings of our friends, and our group started to disintegrate. They were aware of the risks, but did not care too much, and that has been the worst thing they could do. The CTR's, little by little, were seizing the opportunity, altering documents, changing memories, even wiping out somebody's memories, injecting wrong feelings, as easily as if they were showing a movie. Our friends had often stressed that against all that the only possible defense should consist in a firm will, but such a will was missing.

Unfortunately it happened as our friends had forecast, Redda took place (Author: Redda is a negative concept, a disruptive living entity, in opposition to Uredda). As our Uredda was of vital importance, our defection ended in the weakening of our friends also; they told us: "You have not done what we had asked you to do; now we can't withstand the situation any longer; our devices now are ineffective, and in a few days a catastrophe is going to strike, the one we always have warned you against. Now, we do not have enough time to arrange for a sufficient defense program."

The final catastrophe struck in November, 1978, when the CTR's were able to attack our friends, and to enter and destroy most of their bases, even the largest one, the one extending from Pescara to Ancona, and from the central Adriatic sea westward to the central Italy;. I was listening, via wireless, to what was going on some kilometers beyond my feet, the cries, the noises, the orders, in their language which I have never been able to study. They told us: "You are going to see that waters will rise, will be boiling, all over the places where we have built our big base." I was feeling very sick, I was often vomiting, because it was my very world that was collapsing around me. My wife was crying, without fully understanding what was going on.

(Author: during this period the Adriatic sea went mad; see later on for a short description of what happened).

It was a terrific situation; many of our friends flew away, we were keeping in touch just with their central nucleus, that had remained intact, where a few of our friends were barricading themselves to organize a last defense, in order to allow the others to escape. This final phase lasted eight days. Something similar had already taken place one year before, but then it had been possible to prevent it from evolving; this time, on the contrary, it has been a total defeat.

Among the few of us who had remained together, somebody thought of committing suicide, it can't be told the prostration that was pervading our souls; the others simply went away, just like the mice which run away while their ship is sinking. And while fleeing, they lost their memories, of what they had done, and of what they were doing in that moment; they lost the affection they had been feeling for our encounters, so rare, so rich, so happy, as long as it had lasted. And they flew away with anger against me: "You should have let us know that something was to go wrong" they were telling me, but that was obviously preposterous, because I had spent most of my time trying to get them together, and to keep Uredda alive. To me it has been an awful, tormenting moment.

And that has been the end of the story. It went on, on a smaller level, for a while, but our friends were not able to endure. Many of them had died. Sigir and Dimpietro had escaped, but their youngsters did not. The very few W56's that had endured, at the end left, but not immediately; they went away in two groups, one on the 6th, the second one on the 11th of December, 1986. When they left, something broke inside me.

I suffered great losses: I had to sell everything, two buildings belonging to my wife, a couple of agricultural sites, above all I had to sell the large villa I had built on the hill West of Montesilvano, and in doing so I collected no more than one tenth of the global value, because I had to sell everything in a hurry. Nowadays, the villa should be worth some 15 billions of liras, and I got only 1.5 billions from selling it.

To make things worst, I had written two books, one in French, "J'accuse", then another one in Italian, "L'accusa" (Author: both names mean "The Charge") in which I actually charged two of the major world medicine companies with the poisonings they had been making all around the world, and these companies vehemently rejected my charges, but secretly started to plot against me, and in that moment I was very weak, in every sense.

After having sold my large villa, I built a smaller one in Montesilvano, in Luciani street, but our friends, the few ones that were still in touch with us, told me to be very careful, because that was the place where they had destroyed the "Hydra", and it was likely that some negative elements were still surviving. In the Hydra's times, that was country, now it was an over-crowded place. Actually I was not living well inside this new villa; once, when returning home from the Mass, I found Aro, my new dog, a Dalmatian, just dead after the gate. Therefore in a short time I sold also this new villa.

The director of my bank died in a traffic accident, a business consultant who had betrayed me died in another accident, families were disrupted, one inside our group got mad, another went abroad, another started suffering of a nervous disease that prevented him to get out of his house, and so on and so forth. I was submerged with lies, with the most preposterous charges and slanders; three times lawsuits have been pleaded against me, but I have always been declared innocent of any charge.

Any way, I haven't lost my hopes to meet them again: may be some of the W56's shall get back in 2002 or 2003, but not because of us; they have something else to accomplish, I do not know what. May be I'm just deluding myself, because of my strong desire to be again with them; one who has not lived such an experience cannot be able to understand such a desire; he will understand my words, the words of the friend of mine who is helping me to write down these lines, but he cannot feel what we have felt, hear and see what we have heard and seen, the image itself of these beings, their civilization, their sensibility, their infinite goodness, and, above all, their strength; with us, goodness is often considered to be a synonym of weakness, of the desire not to compromise oneself; on the contrary, their goodness is a true strength, because in wickedness there is no strength at all. When one gets angry, and may be even violent, he just shows how feeble he is.

These forty years have been like a jewel to me, both the days of happiness and also the moments of the uppermost fear. In fact, one should speak about two parallel histories, the one regarding the W56's and the other one, related to the enemies of humanity. The enemies of the W56's are very powerful beings, technologically advanced, but their civilization is devoid of any soul whatever. They are people without any future, because they are interested in science only, they are materialists, atheists. They believe strongly in their attitude, which I consider, on the contrary, a sin against our Creator. It may look strange that such a situation takes place, but that is because of the dualism that permeates everything, from the one's self to the whole universe. Our friends have saved us from the evil deeds of their enemies, and in doing so they have also saved our future, but here nobody is going to be aware of that. These extraordinary beings, who belong to the cosmos itself, have come here to help us, not to hurt us. They have not taught us to steal, to kill, to be proud of ourselves, to dream of an empire; their religious attitudes are beyond any imagination; they maintain that what we know about Christ is but an obvious evidence, and of course their knowledge on this subject is much wider that ours.

While writing this book, I believe I am presenting the reader with the information about such great a people, it's a kind of gift from my part, and I hope that when the reading will be over, he will look at the night sky with different eyes, being persuaded that in those distances beyond any imagination, beings live that are friends to all of us.

About our friends

Let me start with some linguistic comments [28]. First than all, their names; of course W56 was not their real name, but it had been chosen according to reasons connected with our reality; 56 of course stands for the year when this history began; the "W" stands for "double victory" (in Italian, the letter "W" is named "Double V"; moreover, the Italian word "vittoria" means "victory"; therefore W = V + V = Victory + Victory). "When you win your enemies – they were telling us – you have still to gain another victory, this time against yourselves, against your pride." Their true name was "Akrij" (The English pronunciation should be something like "Aukree"). Following a suggestion by Mr. Fabio Siciliano, I have found that there is the Sanskrit word "Akhrij", meaning "Sages".

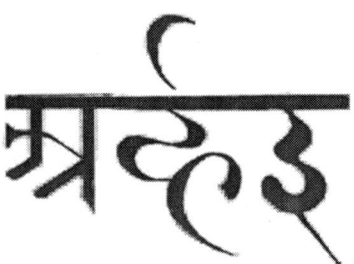

Then, searching in the massive Egyptian Dictionary by Sir Wallis Budge [22], I have found this name also in the Egyptian mythology: it reads "Akriw" (English "Aukreeou"), and the "w" ending is the plural suffix.

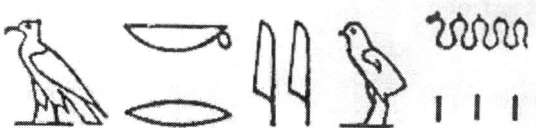

This name refers to some deities who were in existence even before Ra (the supreme Sun god) himself! Even more, they are said to be Ra's ancestors. Any way, it seems that there is but just one quote about these deities, in the 153rd chapter of the Book of the dead, Theban version. Unfortunately, both the books I have with this hieroglyphic text do not contain the 153rd chapter; another one presents only the Italian translation of the Book of the dead, and indeed the 153rd chapter describes the <u>Akriw</u> as <u>Ra</u>'s ancestors, without any detail.

Regarding their – and our – enemies, they are called CTR from the Italian "Contrari" (= "Enemies"); it has been us to choose the names W56 and CTR.

Any way, when our friends chose to be called W56, this name was to be found everywhere in their environments, it was written, in our letters, over most of their objects. Beside it, very often there was a symbol, a kind of hieroglyph, that – they told me – was the name of the galaxy they were from, written in their own language. Actually the W56 were a confederation of many different peoples, so they had different languages, but there was a language common to all the nations, and it was written in some kind of hieroglyphs, or, to say better, ideograms; in Italy, only Emilio and – just a bit – myself had been trained to understand something of this language, while in Germany there were people even able to speak it; it is a very unusual language, with a quite totally periphrastic structure.

They do not have a *Corpus iurum* [31], a formal civil code with which they should comply; they are an highly developed people, so that they are feeling within themselves the way they must behave. Our formalities are unknown to them, they do not have lawyers, courts, or the like; usually they do not quarrel among themselves; when by chance this happens, and that is really unusual, they suffer this situation in their own bodies, they change the color of their skin to a very light white, then they do their best to solve the problem; in this situation, also their machinery goes mad, they say that their machines weep! But, I repeat, this is a very unusual occurrence. To them, doing evil is a plain absurd, they are actually unable to behave in an evil way; they cannot lie, they cannot hurt anybody, nor anything.

They come from a remote galaxy (Author: Please do not take by face value technicalities from Bruno's part: as far as I know, the W56's are from inside our galaxy); the planet where Dimpietro was born orbits in a two-stars system, those people have two suns to give light to their days. Another planet of theirs had a single sun, but a couple of moons.

The W56's history is quite an old one; they originated in a couple of different planets; at a certain moment, these two planets have been destroyed, and the two people emigrated towards a larger planet, and settled there; this happened around one and a half million of years ago. They are saying that their history if like a very long river, that nobody knows where it originates from, but that everybody is aware of where it is ending. At the very start, it looks that they were three different races: the red-haired ones (the "copper-hair people"), the black-haired people (more or less similar to our to-day Indians), the third race had white-greenish hair; by the way, their skin was colored akin their hair. These three races have evolved together with one another, then, as time went by, they started to inbreed among themselves, so that a single new race got born.

In the largest planet of their confederation, some 15 billions of people live. The vegetation in there is different from ours: the greens are lighter, the browns are stronger than in our plants, a chocolate-like color. There is more oxygen in their atmosphere, so that the sky is bluer than ours, and the air has a strong smell.

They feed on earth generated substances; in Dimpietro's planet the soil is generally gray, rich in salts; they breed animals, but they do not feed on them; their animals are different from ours, but not so much. Their plants bear usually very large flowers, with a strong smell; they extract some nourishing liquid from large fruits, similar to our pineapples, but much larger.

They suffer very few illnesses, and they cure themselves in a natural way. For instance, they make use of what they call the "perspiration of plants", which, to me, is more or less the morning dew. They collect this liquid into a large glass, that they place at the center of a table; this table is colored in red and blue, inside a room that is green (Author: colors are meaningful to them). Then they address a speech to this liquid, telling it what is the illness that should be cured, and how this is to be done. Then, they pour the dew into some "ethereal" chalices, glasses that are invisible to bystanders, and they drink this liquid. To be able to see these chalices, one has to use a peculiar kind of spectacles. Once I was admitted to be present at this ceremony, and I was not able to understand what I was seeing, because everybody was apparently drinking nothing from invisible vessels; after having toasted and drunk what appeared to be nothing, they acted as if putting these invisible vessels down on a silver tray; then, after a few moments, these glasses turned to visibility.

They do not clean their bodies using water, as we do; they enter something that looks like a shower, but it is not; then, they are engulfed in vibrations, and you could see a foam emanating from their bodies (Author: There is a yellow light active during this process), and they get totally clean. Their suits are living entities, that adapt themselves to the body they cover (Author: This is a strictly personal process: one cannot wear the clothes of someone else) and protect their owner against a quantity of dangers. Once a month on the average these suits are purified, by putting them inside a peculiar machine for a few minutes.

In their planets, roads are different from ours; when they have to go from one place to another one, they activate a kind of mean of transportation, that emerges from the road itself, and upon which they are able to move to the place they are heading. This mean is available only in urban areas, not in their woods, or in their farms. In these places they have to walk.

From a sexual point of view, they copulate like ourselves; but very often they use a technique of artificial insemination, so that the child gets born inside some kind of machinery.

Young boys get vaccinated against some kinds of illnesses, but their physicians never attempt to alter their genetic system; they say that the genetic system is like a karma, it depends on the past of the person; if one alters it, unpredictable effects may result.

Strangely enough, most of them do smoke; Dimpietro was a real chimney, he liked very much our cigarettes, cigars and pipe tobacco. Any way, usually they smoke some kind of herbs that are profitable to their bodies. Dimpietro was telling me that within our atmosphere a lot of bacilli do dwell, that may be dangerous to our health; many more bacilli, unknown to our physicians, get to our earth inside meteorites, it is not true what our scientists maintain, that high temperatures destroy every form of life; hopefully, most of the bacteria that reach our environment are not able to survive here. Any way, what our friends smoke protects them, in a way or another, from these risks. Another heavy smoker was Itaho; he was living inside a small secondary base under Villa De Riseis, in Pescara, near to the river; he had a small terracotta pipe, and often Giancarlo was taking him some tobacco for it; Itaho liked especially Scottish tobacco. I had presented Itaho with a briar pipe from "Non canta la raganella" (Author: A well known, good, Italian pipe brand), and when at night we were strolling along the sea shore, for kilometers, he was usually smoking his pipe. Itaho was only a bit taller than our usual, some 2 meters 30 cm, so at night he was feeling rather sure that he could go unnoticed. He usually would walk bare-footed, because he liked to touch the grains of sand and the sea water with his soles. Also Dimpietro was often walking together with us, and of course it would have been difficult that he could get unnoticed; so, when he was feeling that there was somebody around, he usually would sit on the sand, so that he could look as a normal man, from a distance.

They breathe oxygen, and they need a slightly higher percentage in the air than we do. Most of the oxides present in our atmosphere are poisonous to them. For this reason, when they have to come to our Earth, they take some substances that enable them to safely metabolize our oxides, and do so about once a week. Their blood is identical to ours, has the same color, only there are much more proteins in their blood; these proteins work, in some way, to eliminate physical scum.

From a physiological point of view, they are roughly identical to us. The main difference consists in their liver, that changes its dimensions and its functions according to the environment. Their liver acts like a sponge, but when its services are not required, it reduces its dimensions, it becomes as large as a fist, it looks as atrophied. When they live here on our Earth, their liver is very active, and works as a filter for the toxins they receive in our environment. Their legs and arms are very strong, while their head is really weak; their brain is somewhat larger than ours, but they maintain that this doesn't automatically imply that they are more intelligent than us. Within the confederation of the W56's, there is a race, whose people have a smaller brain, but they are very intelligent and skilled; for instance, if they look at a man playing a piano, within short they are able not only to play as the man had been doing, but even better, in a creative way.

Most of them have blonde, almost titian, hair. On the contrary, a few of them had black hair, deep black with some bluish hue. Among their youngsters, some have very short hair, in the German way. Their hair is thicker than ours, and the same is with their eyelashes, that are rather long. Some, not many, are growing a beard. Their beard is very stiff; even when they shave themselves, you could feel their beard under their skin; to get shaved, they do not use blades, but a kind of instrument that cuts their beard at skin level. As usual, Dimpietro was behaving more like ourselves, he was using a straight razor, not a safety one, but one of the kind our fathers have been using; only, because of the peculiar stiffness of his beard, he has had to build a special razor, with a blade made of some specific metal. Once I had asked Sigir on the reasons why Dimpietro was so much similar to us, and he confirmed that they also were wondering about that: "He has been this way since when we have met him; at times we wonder whether he is actually a terrestrial, may be he left your planet when very young; indeed, his habits are very similar to yours".

Their hands are thin, just like their bodies; I have never seen a fat man among them, nor are they skin and bones: their bodies are just lean in a right way. Their fingers are long, and they're ending with nails that are a bit longer than ours, some four or five millimeters longer.

Their eyes are gray, blue; some had black eyes, but again with a bluish hue. Their iris is significantly larger than ours. Moreover, I've never seen them blinking, batting their eyelids.

The smell of their bodies is reminding the wheat, it often made me think of the kneading-through that was frequent in peasants' houses many years ago. Then, when three or more of them are gathering together, another smell can be felt, that of a sylvan pine-tree. They love perfumes, theirs are a bit bitterer than ours.

Our friends were exhibiting a very calm attitude, I admire them for that; their face was always quiet, but not in a pretended way, their serenity was coming from within themselves; then, there was always a faint smile on their lips; I have never seen them angry, nor frowning; their voices were very similar to one another, calm, deeply resonant, likeble; they were never speaking in a loud voice, only, when they wanted to attract your attention, they just modulated the words in a slightly different way. Their cool attitude was reflecting their innate strength and maturity.

They are wearing very simple clothes. Directly over their skin, they have a very tight overall; it looks to be made of nylon, or of synthetic cotton, but it is neither; they call it their "second skin", and it is intended to protect their body against both warmth and cold; moreover, it is transpiring, and while preventing toxins from reaching their skin it has also the function of sending out body impurities. It is flesh-colored, and very soft. These second skins have to be purified every tenth day, on the average.

Over this "skin", they are wearing an one-piece suit, tight at the neck, wrists and ankles; this suit ends with a pair of shoes over their feet; these shoes are not different objects, they are a single piece with the suit. They are like small boots, their soles are some 5 centimeters thick, and look as being made in rubber, but it isn't so, I was told. The soles are designed so that they may have a firm grip over every king of ground.

These suits are colored in every hue of gray, green, red and blue; I have never seen yellow, violet or orange colors, at least not a strong orange. At the neck, the suit has a rigid and tight collar, often bordered in a light orange. They also wear white suits, specifically in occasion of religious activities. It looks that they are attaching a lot of importance to the color of their suits, in accordance with the peculiar activity they are involved in. For instance, when they have to get together to take important decisions, they typically wear a suit of a translucent, almost pearly color; they were calling it "the wandering colors", because actually it looks as if different pearly hues are moving all over their bodies.

For a few minutes every day, they perform a kind of keep-fit exercises, with very slow movements, and they have taught me some of these practices. Other times, they gather in groups of three, four people, sit on the floor, and sing strange songs; they do so when some discord had grown among themselves, in order to settle things down again. They also pray, on a regular basis.

To have their meals, they gather together, sitting around a green table, bordered with a narrow orange line. Each one of them has in front a kind of metallic tray, sub-divided into smaller compartments; inside each section, different kinds of food, mainly something that looks like small pills. They are using a kind of silverware to get the pills and eat them. They told me that, on the average, they evacuate just once a week, because their food does not leave any residue. "We must drink a lot of water", and I thought it should be plain common water; not at all! They are putting a few "silver almonds" inside a kind of crystal bottle, then they pour a liquid that to me looked like simple water, but it wasn't. "This is an artificial compound; there is water in it, but also all the minerals our organism needs; it is mainly rain water or also water collected inside the clouds, because there it is very pure".

Our friends were also feeding on the fruit we were collecting for them, but not directly: they were extracting their vital particles, and using them to make their pills; once I was offered a kind of greenish jam, and was told that just a minimum quantity of it would suffice for a couple of days, without any additional food. "It is not a stimulating food – they said – one that only deceives your organism, and later you must pay the consequences, it is not an artificial meal, it is totally natural and effective. The only problem is that it is extremely bitter; would you like to test it?" Of course I would, and at once I felt satiated, but not filled up. Giancarlo did not want to test it: "Cos'è 'sta zezzità?" (Author: An almost untranslatable vernacular expression, roughly: "What's this ugly mush?").

They are sleeping just two or three hours per night; again, Dimpietro was different, because he was able, at times, to sleep even a whole day long.

Dimpietro was a bit different, actually near to our attitudes, and his friends were often joking on him, saying that he had acquired terrestrial habits. And he was laughing at them. Often I was telling him "You have been born here, you belong to our race", and he went on laughing. I have already told you that he had learnt many Italian dialects, and was able to speak with different accents, even to tell jokes, to the amusement of bystanders. He was also used to eat much like we do, pasta, fish, seldom meat, bread, vegetables; he also liked to drink, especially wine; he was able to drink wine pouring it from the flask directly into his mouth, in a way that only our older peasants are able to remember, and he was doing so just as a joke. He had been living so many years over our Earth that his organism had grown used to our habits. Once he told me that, to adapt to our environment, he had had to undergo some minor surgery, something to his lungs, I remember, so that he could be perfectly fit to our planet.

Dimpietro and Sajù were often quarrelling but mostly as a matter of a joke. Once, for instance, there was something to be done outside, and Sajù told Dimpietro: "You go outside, because you are better fit to the habits here, and so if you meet somebody you'll be able to get out of trouble" "Yeah, if I meet somebody he's going to get a stroke at seeing a man as tall as I am; you go, you that are almost a dwarf to these people!" They were spiteful to each other, and often their jokes were rather harsh: once Dimpietro got really angry at something Sajù was mumbling, and so he first extinguished his cigar on his bald head, then he took him in his arms, and spat over there in order to relieve the burn!

Both Dimpietro and Sajù have been living on this Earth since very long ago; in the 17th century, for instance, they have been living for 15 years in the Persian Gulf, mingling with the local pirates, to study their environment and habits. They found, with amazement, that most of the pirates activity was promoted by various legal governments. Once they have been captured by English troops; Dimpietro pretended to be a magician, and in some way they were able to get free. Sajù told me that in those years, he looked really like an ape, and that he was very fond of bananas; Dimpietro had been presented with a parrot, and he kept this bird for more than 150 years; as it felt ill, he took it to be visited by some physicians inside a space ship; unfortunately, their response has been that the parrot was just suffering from aging, and that they could do nothing with that.

Dimpietro was loving so much red melons, and so we often were taking some very big melons to him. Usually, then, he would sit in a corner, and cut them carefully into small slices. Once we had prepared a really big melon, and he drew it in the usual way; there were Sajù, Sigir and Romolo with him; "Are you happy?" I asked him; "Of course, we are pleased of much the same that pleases you." Then, we later were told, Dimpietro sat as usual, and started cutting the melon into thin slices, lining them on the floor before eating them, and in the meantime Sajù, begun stealing away secretly, then eating, some of them; Dimpietro did notice that something was going wrong, because he had been counting the slices as long as he was cutting them, and he was a bit perplexed; but when he saw the small hand of Sajù that was appearing to grab another slice, he bit it, and a shriek was heard!

He had told me that he had been in the United Kingdom, that he had visited Shakespeare's house, many museums; "How have you been able to, as tall as you are?" "Well, I went there when there was nobody else". He was very interested in Goethe, Kant, Nietsche, and many other philosophers, but he was not able to understand philosophy itself, because, he used to say, "It leaves you with many questions and no answer".

He was very fond of music, he liked to play the violin; I have seen this instrument, and asked him whether it was a Stradivari (Author: Stradivari has been an exceptionally competent Italian violins builder, and the instruments he made are famed all over the world for their beauty of tone and their design), but Dimpietro answered that he himself had built that violin, in his home planet. He would not accept to get disturbed by anyone while he was playing his violin.

While Dimpietro was exceptionally tall, Sajù was quite short; when Sajù was coming to meet us, Giancarlo often was saying "Look, the dwarf is arriving!". His voice was strident. The short people among our friends, on the contrary, have a normal voice. Sajù had also a bald head. "When I was a boy – Sajù used to tell – I had been predicted that I was to become as tall as these ones, but something went wrong!". He too liked very much to joke, even on himself. Dimpietro often was stating "It has been your malice that has prevented your growth!"

Then there were the youngsters, I have seen many of them; actually it is difficult to guess the real age of our friends, because, for instance, they do not develop any wrinkle with age, their skin is always very smooth and fresh. I had asked them how they look like when they approach their physical end, and was answered that they are actually able to contrast aging, and that they die when this ability cannot be maintained any longer.

To them, religion is at the base of everything. They see God even in the smallest insect, and pretend that the whole universe has been created, and that God is everywhere inside the cosmos. Their religion is not as full of rituality as our cults are, to them it is just a deep feeling, that doesn't need any appearances. Typically they perform their rites thrice during their year, and, as an exception, in the occurrences when they need a stronger support. In such instances, they gather into a circle, and they sing something very similar to Buddhist chants; in the center of the circle there is a crystal column, and inside it a metal plaque, it is not gold nor silver, but something in between. This plaque rises as the rite goes on. When it gets to a certain height, they start asking God for what they desire; if they are to be fulfilled, the plaque gives a flash, then falls down, elsewhere it falls down without any other effect. Once I have been shown such a rite, and it looked very simple, even childish, but it was not, because of their strong religious feeling that was making appearances useless.

I was enchanted by their devices, although I am not a technician, nor did I care too much about their technicalities, and when they told me that they were about to leave, I asked them to let me keep some of their gadgets; unfortunately, they told me that it would have been useless, because the whole global structure of their bases was required in order that the smallest equipment could work. I was not so glad, but they went on: "We could even leave a whole base to you, with the whole of its apparatuses, but you would not be able to operate it. Moreover, were you to call some of your scientists for help, you would get robbed of anything in a short time, because each scientist would try to spruce up to the eyes of his own government."

Giancarlo was very interested in technicalities, and he was trying to develop new inventions based on the technology of our friends. They were telling me that this *ayuno* (this word means "brother" in their language) was actually an inventor, that he had good capabilities, but that at that moment there were no conditions, on our Earth, that could allow him to develop any project. Once he decided to try to build a flying saucer, and our friends told him to forget it, because he was not in the condition of getting to any result, nor was anybody else on this Earth.

They are very humble, not in the way we may find among our environments, people that pretend to be humble just to conceal their innate pride; they were actually humble, because, they were saying, humility does not mean weakness; never did they look proud of their capabilities, because, they were saying, that was not a merit from their part, but just an accident in the history of the universe. They are always quiet and smiling. Particularly, Dimpietro, Gallarate and Siderius were showing a strong personality, but nevertheless they always were very friendly. Once I asked Dimpietro whether he was to be considered some kind of an holy person, and he fooled at me.

They had women with them, I've seen at least six women inside their bases; they were really beautiful, and you could feel a strong sense of femininity emanating from them, but not in an erotic way; Giancarlo, fool as usual, once got even in love with one of them, and her friends were mocking him! They got no problem with nudity, in the sense that they attributed no meaning, no importance, to being seen naked, both men and women. Probably it was only as a mean of respect to us that they usually were avoiding to go around without any dress, but actually they were not paying too much attention at that.

Of course families are formed, and children are born, even here on our Earth. They told me that when the youngsters are only a very few years old, they must receive the memories of their people, and this operation is made through strange machines that actually load their memory with the history of their race, and their culture. The operation involved wearing a kind of silver cap, and to feed some small cylinders into the main machine, cylinders that were almost as little as one of our AAA cells. Each cylinder contains a portion of the global memory.

Their mission over our Earth was to look after this planet, and above all after our safety, without interfering. And they told me that, since 1956, in at least two occasions they have had to prevent an atomic war, and they did so by transmuting the fissionable metals inside war-heads into lighter substances, so that no nuclear reaction could take place. Another unexpected thing I was told is that they are always receiving messages, mainly via wireless, from Earth people who would like to get in contact with them, but they are of no use, I had been told, because it up to them to chose the people they want to contact. They told me also of the messages sent into space by our powers, USA and Soviet Union mainly, but were saying that "It's a waste of time, because these messages should have to travel for so long a time, and so they are totally useless. We know everything about you, do not need any specific message from your part".

The CTR's are quite at opposite: they adore science only, therefore they are very cool, when they believe it to be the case, they destroy people without the least hesitation. They have a kind of "scientific ethic", and this name explains their behaviour. It had been us to chose the name CTR, from the Italian word "Contrari" (= "enemies"); our friends used to call them "Our enemy brothers". Most of the times, while the CTR's were moving in our environment, they were driving dark blue Mercedes, the diplomatic model, i.e. a very large car, with four of them inside it; once Dimpietro told me: "When you chance to do so, try to measure a Mercedes used by the CTR's: you'll find that it is a span longer, and a span wider than the usual model". On the sides of the car, I had been able to find many small meshes; Dimpietro had told me that they were a kind of offensive device. Another weapon was on the back of the car: it looked like a small circular metallic plate, with many small holes all around; through these holes a gas was supposed to get out, that was very dangerous to human beings. The CTR's too look like normal men, were usually bald, and their bodies were giving out a smell like tar.

Our friends told us that there were also other people coming to our Earth from outside; some of these were something in between a man and an ape, like beasts in their path of evolution, and we have been shown pictures of these beings; but actually Darwin evolution cannot take place: an ape will be forever an ape, a man a man; never the ape will transform itself into a man.

When our scientists have been saying that UFO's do not exist, our friends didn't feel hurt from that, nor have they said that our science people weren't but a kind of fools; they have respected their opinions, just saying that they were feeling rather sad because of our mistake.

Wars do exist, one is the war between them and the CTR's: the CTR's have come to our Earth to harm; they do not abduct people, this phenomenon that to day gets so large an echo in UFO's books cannot be charged to them. While this war was taking place, almost nobody on our Earth did notice it: our earthlings were looking at TV sets inside their homes, they were reading newspapers, they were going to dance, but never did they pay any attention to the sky over them. Here on Earth nobody is able to notice the changes that are taking places all around us. Our friends were saying that they had acted as the positive bacteria on our planet: we did not notice them, but they in the meanwhile have acted for the good of our Earth. Hopefully, before leaving, our friends have arranged things so that the CTR's will not be able to trouble us for several years to come. Any way, they may try to do something, and from our point of view, a little thing by the CTR's converts into a big deal to us!

Some of Their Teachings

They call our Earth "The Universal Center for Redemption", because they pretend that the souls who get incarnated over this planet are the ones who have yet to fulfil their evolution. That's why here there are so many sufferings, but at the same time also so much success may be found. Moreover Earth is one of the most beautiful and most complete planets. Its history is much longer than the one we acquainted with. It has seen many more civilizations than our books tell.

The Southern America has been the cradle of a very advanced civilization, well before Egyptians. They had strong ethics, but at the same time they had achieved a very high level in science and technology, to the point that they were able to operate their devices via mental control. Ambition, vanity, will of power did strike even there, unhappily, and everything got ruined; such a thing did not happen just once, but many times. Our friends have shown me maps of the surface of the Earth even before the time of dinosaurs. The so-called Stone Age does not mark the beginning of man's evolution; on the contrary it is the end of a civilization, after it had destroyed itself, and the survivors had to start again from scratch.

Our friends were often saying that our scientists behave in a strange way: they often get to significant results, but then they are unable to go on, because they impose limits to themselves. How may it be possible that in the universe, in the infinite, limitations may exist? Our scientists behave as a glazier that uses a diamond to do his job, that's to say that uses a much more valuable item to work on lower substances. Our best engineers are a kind of mystics of technology, they put their science on the top of anything, and in so doing they block away any possibility of further evolution of their discoveries; our friends were telling that they have also invisible machinery, the simple idea of which would be preposterous to our scientists.

They were also saying that doing good is not only a duty, but a must. Were men capable to do good sincerely, without hidden purposes, no sadness could be found among them, and that might be the start of a new Renaissance. And such an effort would not cost anything; there is no need to spend money, to buy books, to attend courses, in order to enlighten oneself; it suffices the will to do so.

About the free will, our friends maintain that it exists, but there is also an universal condition, because of which everything that has to come is already stated, and when even the simplest thing happens, it fixes a lot of future events, up to a future very far away. Therefore, to make use of the possibility of free will, one must be well aware of the environment, of all the future consequences of his own action, and moreover one must be able to limit oneself to what is strictly necessary, in order to prevent the widening of the range of the effects that will arise because of his action. But usually most people are not able to behave in this way, therefore they only believe to be free in their decisions, but actually they aren't, because they are unable to understand the constraints imposed upon them by their own inability.

The planets of the W56's were distributed in part in our galaxy, and in part in another galaxy, that they were calling the "Blue galaxy" [23] because of the color of some filaments inside it. Once I had asked our friends "What is the situation up there? Are there you only, are there other people? Here we know nothing about aliens". "We love your Earth, it is wonderful; among the planets we know, there is some that looks a bit like your Earth, but none of them is as dear to us as it is. Unfortunately, man is now ruining his planet; from heights we may perceive something like a set of spots, like in the head of a man who is getting bald. For what concerns civilizations, I must say that over your Earth there is not so much of it. You've got inspired people, saints, teachers, mystics, it's a bit as if the Universal Center, the Supreme Entity, has paid Himself a peculiar attention to your planet; by the way, you know only a part of your own history, and have only vague feelings about what is missing; I may tell you that also in the antiquity you are totally unaware of, God has done so much good to your Earth. But you have never been able to profit of His help. For instance, thinking about Buddhism, you have not yet understood that it is a mean to open one's mind to higher levels, because once you've got the Nirvana, your mind is quiet, calm, and it is easier to it to open itself to the infinity. All the technicalities you perceive within the various religions are actually intended to make your bodies and your mind open to the light coming from above. This light comes to everybody, but most do not take it into any care".

"There are other people besides us, at various levels of civilization, but the man is universal: you may find small variants from one race to another, even among ourselves you have seen very tall persons (myself for instance), and very small ones; there may be differences in the skin color, there are people whose flesh is almost transparent, but, I repeat, almost every civilization is made up on men. In some planets there are animals that look like a man, a bit as with your apes; but remember that one million years from now, apes will remain apes; the theory according to which man is an evolution from apes is totally false. Such a physical evolution simply cannot take place, and your biologists who believe to have reconstructed the path from, say, the *eohippus* up to your horse are totally wrong. A race, whatever race, never gets modified if not by an accident, and when this happens usually the mutation is a degenerative one".

"Our people has lived many ups and downs, in several instances they were on the verge of annihilating themselves, but hopefully we have always been able to overcome the risk. To many other peoples, techniques and science are held as the most important things, so that they forget that people are souls, above all; if one forgets this simple concept, he is going to face serious risks. It has actually happened even to us, but we have saved the situation. We have to thank the forces that make up the universe, we have found a great help in them, and now we are at a very high level in our path of evolution, not an evolution in our bodies (or at least just a bit of it), but above all in our souls. There are several different races within our confederation, and not all of them are at the same level; the ones who have resorted to technicalities only have degenerated, or even they have died as a people, because if you base all of your society on science, you must take great care to prevent anything from going wrong, and, if it happens so, you're lost. Unfortunately, you are going this way, and you are going to face serious problems in your near future. Moreover, there is also another problem with you: money. You could solve many of your environmental difficulties, but the great economic powers prevent such solutions from being adopted".

That must be true because, aside from my own experiences, I actually had personal acquaintance with a couple of guys who, upon directives by our friends, went to Namibia, in Southern Africa to develop a new kind of motor, and, after a few months, nothing has any longer been heard from them.

"As man is universal, albeit some secondary differences, all of them breathe oxygen, because all inhabited planets have oxygen in their atmosphere, of course with differences among them; we are very good to adapt from a place to another, so it has not been really a problem to come here on your Earth. As you know, Dimpietro (here Ljufur was speaking) has adapted himself so well to your environment that he has also adopted your habits and your tastes!" Dimpietro was very fond of our way of living, of course he was not indulging in sinful or dissolute activities (Author: As a smoker, I am sure that most Americans will not forgive his habit of smoking); he was smoking cigars, was drinking a lot of coffee (Italian style!), was appreciating every kind of music, was fond of looking at youngsters dancing, and he too was often dancing; on the contrary, he did not like our moral filth; he was not at ease with our habit of putting sex above everything: sex of course is a necessary part of the life of everybody, but we often tend to put it in the center of our life, and that's wrong.

Dimpietro and the W56's, while appreciating our way of life, were also tolerating some of our bad habits, until they were not in conflict with the environment, or with ourselves. For instance, Dimpietro was a strong smoker, but he was assuming something that would purify again his lungs in a matter of hours, so, from his point of view, smoking was not actually a degenerative activity. Very seldom he was eating meat, just for the sake of trying a new experience, even being aware that meat would have been harmful to his organism, but, again, he immediately would take something to prevent any damage. Aside from the risk of being poisoned by meat, the W56's usually would not eat any meat because they love animals, every kind of them, and so they cannot tolerate thinking of eating any animal flesh, in the same way we could not think of eating any human flesh. Meat is a danger also to us; I have seen beasts killed in a slaughter-house, and I know that, whatever precautions may be taken, animals are terrorized, and when we eat their meat, we eat also some of the evil seeds engendered by their terror; I can't imagine eating meat from animals killed according to Hebrew or Muslim habits, that enhance the frightening in the poor victims.

According to them, death is not an actual end of a process of life; our religions affirm that the soul survives when a man dies, but our friends were maintaining that something survives also when an animal, or a plant, dies, and that this surviving "part" is higher, the stricter the vicinity of the living being to man. These entities organize themselves, and play an important role in the future of the species. Reproduction is not due only to seeds, it is also influenced by this kind of entities; behind every physical phenomenon, there is an invisible action that we, men of this Earth, are unable to notice.

One could wonder why they do not act in some way to relieve us from our miseries, both physical and moral; actually that is a task impossible even to them: our civilization is too much different from theirs, and in order to change our way of thinking into theirs, they ought to ravish our identity, so that it should get destroyed. That can't be done on a large scale, because we would get annihilated, in the way South America people reacted to *Conquistadores*. Their idea was to gather a small group of enlightened people inside my villa, and to train them so that they could behave as trainers to their fellows, and so, little by little, our attitudes could change, but unfortunately this design went into pieces, mainly because of the inability of most of us to cope with it.

They were stating that our civilization is ruining towards madness: we are destroying our environments, we are destroying our green, our animals; that little that our science is able to pick up may be of some use in treating some of our problems, but often it's the cause of new problems, of new illnesses. What we are doing in the fields of medicine, of genetics, of physics is but the topmost madness, and in the next tens of years we are going to pay the consequences of our insane attitudes, and our friends were quite aware of that, and were very sad for that.

Our friends have come here to do their best in order to help us; we were not in the position of telling them what they had to do, because we have only some little hint about their designs, and I am sure that if anything else may be done, in the future they will comply; our friends told me that ours is one of the best planets inside the whole galaxy; there are other fine places here and there, but the Earth is really peculiar, from a moral, a psychic, a human point of view. It's very pleasing to look at our Earth, it's a satisfying feeling looking at it from a distance. Our trees give us ions, balms, they purify our atmosphere, and we despise them, and destroy them just to get their wood, just as if we would kill someone to dispose of his body.

Roughly speaking, according to them, the number of contacts between them and humans is something around 5% of the total number of sightings from our part.

Once I asked them if they knew anything about the Little People, what we call elves, gnomes and the like. "Of course they do exist – I was answered – but they conceal themselves because of fear; they are 35÷40 centimeters tall; they are very nice, merry, joker, generally speaking they are simple people. During your Middle Ages, they were feeling less endangered by your people, so they were noticed more frequently than to day. They are harmless. There are also space people as tall as that".

Some remarks from my own part (Stefano)

The history of Amicizia has been always passed over in silence, although here and there at times there have been exceptions, among them very important are the books written by the Consul Alberto Perego; he dared to quote it explicitly, and to present some interesting pictures. But, at his time, the Italian diplomat had not been taken seriously, but hopefully to-day he is getting appraised again, *post mortem*. Something else appeared, *passim*, in the books presented in the Bibliography, but always in a concealed way. This book is probably the very first one to meet this topic fairly and squarely (shortly a second one will appear, signed by Prof. Paolo Di Girolamo, and may be even a third one, written by a professor in the university of Padova). By sure the subject is hot, under many points of view, therefore it's difficult to decide what to tell, taking care not to involve, unintentionally, persons who prefer to remain concealed.

While writing down Bruno's tales, I realized that his point of view is not exhaustive, therefore I believe that something *a latere* has to be added, and I'm going to do just that, referring to information I had got from Hans and from other people who do not want that even their name be written down.

First than all, the mass contact phenomenon called Amicizia (and similar names in other tongues) has not been restricted to Italy, although our country has been, in a certain way, the nucleus of it, followed by Germany. In Italy there were many persons referring to Bruno, and a small number of stand-alone ones; I have known some fifty persons in Italy, a couple of French people, four Swiss, one Austrian, one Russian lady (from Sibir!), and some twenty Germans who, in a way or another, were in touch with the same presumed alien entities. At a guess, I may suppose in one hundred the number of people involved. It looks that something has taken place also in Australia (Bruno himself refers that) and in Argentina (I have found vague traces while I have spent a long time in that country, because of my job). The phenomenon has been really a noteworthy one, I'd say the hugest case of mass contact, both from the point of view of its duration, and from its geographic extension.

The setting was the diatribe between the W56's ad the CTR's, a very concrete struggle, *with casualties from both sides*, both among the aliens from the two groups, and among their Earth supporters. The role of earthlings was first than all a logistic one. Bruno has told of the trucks full of fruits he had to organize twice a month; they had to provide other items, typically metals, often ones not so easy to be found. At times the presence of earthling was required, when it looked that the superior technology of our friends was not able to cope with something (I'll tell something about that later on). In return the W56's were giving a huge amount of information on many topics, from philosophy to religion, from biology to physics and engineering, and, from time to time, they were also participating to the heavy expenses earthlings were usually affording (most of the times, their supports were at joke levels).

Any way, notwithstanding the scientific gap between them and their friends from Earth, the W56's weren't acting as superior beings; on the contrary they were extremely friendly, almost brothers (whence the name that so many persons have given them, independently the ones from the others). The W56's were appreciating the best of life, often they were living within our society (at almost all levels), they were loving well eating, well drinking, to have a good time, to smoke (I believe it's the only case in the history of contactism, but in these very last days I have been told of another alien, he too a strong smoker), they were fond of music, of arts (although at times they weren't sharing our points of view).

The were quite human-like, with some secondary differences: very stiff beard and hair, longer hands; the most conspicuous difference was in their heights, from six meters to some ten cm. Obviously only "normal" people, among them, dared to enter our community, but there were many exceptions, some of them in Bruno's tales.

Long distance contacts were taking place mainly via radio broadcast: these guys were able to send a message to a particular receiver, wiping away the normal broadcast, with an incredible selectivity: if one had put two receivers one beside the other, both tuned on the same frequency, only the pre-selected one was to receive the friends' message, while the other was going on, as if nothing was taking place! Once, at Bruno's, I got a message from a transistor set (it wasn't a radio, but a portable disk player, therefore without a tuning section); first I cut the power cable, then I opened it and cut one of the loudspeaker threads, and nothing changed! The loudspeaker was going on, as if everything was OK (please remember a sentence by Williamson quoted in the first chapter): usually a tuning unit was not necessary, a loudspeaker was enough.

The W56's were able to listen, from far away, at what their interlocutors were saying, and this was the way back. Actually, there were able to use no physical channel at all, and to have their voices heard directly where their interlocutors were; but they were maintaining that they preferred to avoid doing so, because they had felt at times some fear by the earthlings involved.

In alternative, the rudimentary e-mail available at those times was used (no Windows, no PC's, typically IBM 2741 teletypewriters, connected to a remote computer via a modem), or, more trivially, telephones, composing numbers 35 digits long on the average, when the telephone switches in use had been designed to handle up to 11 digits, and finding no charge from the telephone company. Not so often something similar to telepathy was used, most times thanks to "anie" (see later), but these were an exception.

Amicizia has hosted a lot of technology, a lot of research from our part, under their guidelines, and some help, from time to time. Now I am going to give readers some scent of all that, and I start letting Hans speak:

In the following, the electrical schemes of some circuits I designed to test some of the ideas of our friends; of course, vacuum tubes were employed, together with the discrete electronic technology of those times; I don't know whether to-day integrated electronic would permit something like that: standardization makes unusual circuits difficult to be made. All these objects are power oscillators, although a bit unusual, ranging from very low to rather high (for the times) frequencies, and the unusual design of the power circuits (due to the characteristics of the device to drive) would be difficult to implement with integrated electronics.

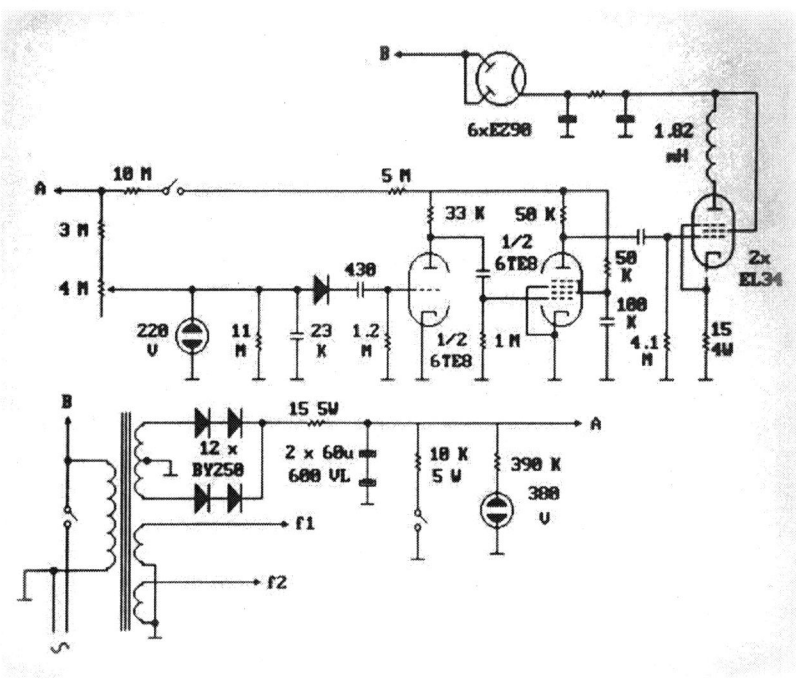

Regarding this diagram, I'd suggest to the few who are still understanding something of discrete electronics, to pay attention to the 12 (!) silicon diodes in one of the anodic circuits, and the 6 (!) EZ90 tubes for the other one, and to the necessity of designing an explicit way to discharge anodic supply when the set was off (the switch just before the neon lamp). The work tension of the electrolytic condensers may give an idea of the anodic tension, and therefore why a switch was needed to discharge them...

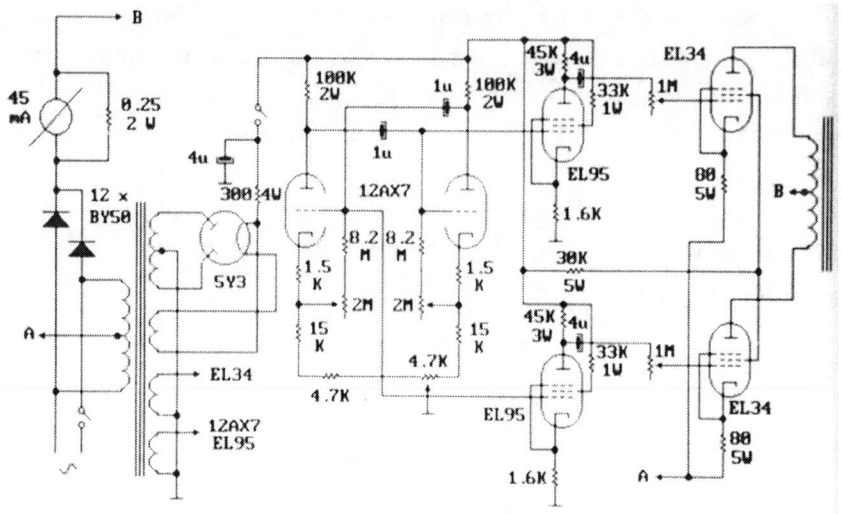

Here we have a trivial relaxation oscillator (who is willing, and is able to, may compute its frequency; any way it's easily guessable from the values of the components), based on the two sections of a 12AX7 tube, that, via a couple of EL95 (!) drives a push-pull circuit based on two EL34. Once again, 12 silicon diodes to give power to this final stage; in this case, anodic tension is not as high as in the previous example. Who is willing, and is able to, may compute the current measured by the ammeter on the top left, with a 0.25 ohm, two watts shunt in parallel...

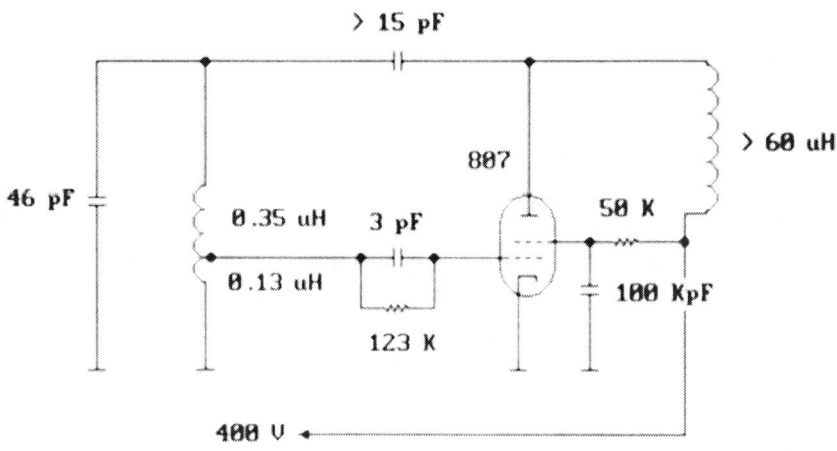

Stefano again, with an oscillator using an 807; this circuit has been part of my doctoral dissertation; the real thing was using two 807's in parallel, and at times their grids were getting red-hot, because of the power!

The following diagram is a computer simulation of the electrical field generated by a particular device; it has been made via a program running on an old PDP 11/70, making use of the poor graphic technologies then available.

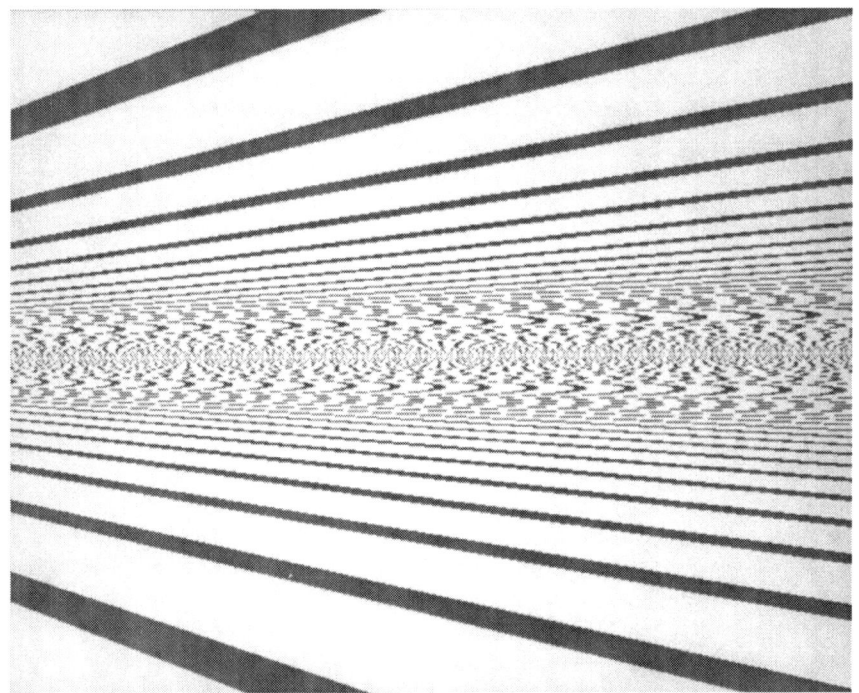

Interesting enough, we may see here an example of fractal structures *ante litteram*! The center of the beam looks very much like a Julia Set.

I have presented these objects so that the reader may get an idea of how deep has been the work made in cooperation with our friends, trying to study whatever they were telling us, often trying to actually build devices to implement some new ideas, or making use of the most up to date technologies available to study them from a theoretical point of view. For instance, this differential equation is strongly related to the fields generated by a stationary flying saucer:

$$\frac{dy}{dx} = \frac{tg\varphi + \rho \cdot \dfrac{d\varphi}{d\rho}}{1 - \rho \cdot \dfrac{d\varphi}{d\rho} \cdot tg\varphi} =$$

$$= \frac{\rho + \dfrac{d\rho}{d\varphi} \cdot tg\varphi}{\dfrac{d\rho}{d\varphi} - \rho \cdot tg\varphi}$$

A mathematical item not easy at all to be dealt with! By the way, the following diagram shows one of the infinite integrals of this equation.

Of course, it has been computed using a finite differences method, because that equation cannot be integrated using the standard tools offered by conventional Calculus. And, I must say, that particular equation is but a gross approximation of the actual thing, because in the reality obviously the generated fields tend to zero with the increase of the distance, and that equation doesn't take that into any account. Any way this diagram explains why, often, an UFO is seen as if surrounded by concentric halos (pay attention to the accumulation of force lines along circular patterns), particularly if the witness wears Polaroid spectacles.

This drawing has been made by one of our friends:

It does not show too many details (its original is a drawing made with white chalk on a black sheet of paper); I present it here because of two curious mistakes in Italian language: the word "Torzione" is misspelt (it should be "Torsione"); in the same way "Bobbina" is wrong (it should be "Bobina"). As, within Europe, Amicizia was gravitating mainly on central Italy, often our friends were speaking an Italian heavily corrupted by the local dialect (the "bobbina").

Of course, their technology is really far beyond our understanding. In the 60's we were gradually moving from electronic tubes to transistors, first, then to hybrid circuits (the IBM 370 was an example of that), and, little by little, to integrated circuits. I graduated in '69 discussing a thesis on high frequency power oscillators (strangely enough, isnt't it?); just imagine what an electronic engineer like me could have done if presented with an integrated circuit, let's say a Pentium, in the 60's! I couldn't have the least possibility to understand what the hell that object was for. And I am speaking of only 40 years ago. Just imagine what we can do with a technology so much more advanced as theirs. Nevertheless, something might be perceived, although not fully understood, and people from an half of Europe has worked oh these concepts. One could be surprised to know that one among the to-day most classic amplifier circuits in the TLC environment has been designed following information coming from outside (Hans too has made a remark on that). Do please allow me to conceal which peculiar circuit I'm speaking about: it would bee too easy, to experts, to find the name of the man who has, supposedly, invented it. As I chanced to get acquainted with him, I must recall that he was an exquisite person, that he had accepted the situation with a really open mind, a situation difficult to be expected from his countrymen, as German people are widely supposed to be four-square persons.

Now it's Hans again to speak:

Speaking about electronics, our semi-conductor devices have been born on a germanium based technology, and later migrated toward a silicon based one, because of a better thermal stability. The W56's, on the contrary, are still adherent to germanium, but they do not use it as a semi-conductor, on the contrary they look for an almost absolute purity. The polarizations we get in an electrostatic way, are obtained by them via a displacement of atoms inside a crystal lattice.

They use a lot of mercury, in most of their applications, so that earthlings trying to imitate them were to cope with this metal: *horresco referens* a device based on a nitrogen plasma generator, with mercury pipes, that got solid thanks to liquid air envelops! Useless to remark that mercury is an expensive object, not so easy to be found in so great quantities, moreover a toxic substance; but we have been working also with asbestos, with radio-active compounds (radium, barium strontium niobate), and with hyper-voltages generators (beyond 1 MVolt), so that we were used to be careful.

They were also making a wide use of iron, almost all of their scouts are made in iron (that's why they're so heavy: an Adamski-like bell weighs between 15 and 20 tons!); not steel, but pure iron, although in allotropic states different from those we are acquainted with, crystallized into a regular lattice (our iron, just like all the metals we make use of, is an amorphous substance). Under such conditions, iron may an almost null thermal conductivity (!), may be made locally transparent (if one likes, even in a mono-directional way), and may become an electronic semi-conductor.

Some Conversations

Now, I'd like to present the reader with a very strange conversation, that has taken place between one of the W56's and a German gentleman; as usual, talks have been in German, I had translated them into Russian, then into Italian, now into English. In the following transcription, the German's sentences will be written in italics.

How many pieces are contained within a box of matches? One, one thousand, a million?

I believe one hundred, when the box is full.

Why not one million?

They could not fit into. Moreover, they would cost much more than the usual price of a box.

But could you find a million of matches in a box you have bought believing that it contained only one hundred of them?

They could not fit into it.

You might also fill a matchbox with billions of pieces, had you the necessary technology.

The ones who actually manufacture matchboxes aren't able to put inside them more than a hundred pieces.

You get always limited by your environment: do please imagine that it may be possible to put a billion of matches inside a 100-pieces box.

OK. So what?

Should you get astonished in finding so many?

No, should I be aware that's possible. Of course I'd do, as I am sure that it is not possible.

But I ensure you that it may be done. So, should you get astonished?

I would any way, because such a thing has never happened to me.

Limitations as usual. If it never happened to you, that doesn't imply that it's impossible.

At least, it's very improbable. I smoke a pipe, so I'm used to open matchboxes, and I never chanced to find one with a billion matches inside.

So, you should get astonished.

Of course I would.

But the universe is made up with anomalous situations; it's them which carry information (Author: That's the Shannon's Theorem); if all the matchboxes were to hold one hundred pieces, nothing new could ever take place.

Were my company to build transmission equipment whose main frequencies were not constant, we should quickly face bankruptcy.

Not really, it would be enough to build peripheral systems able to adapt to single frequencies, just as it should be enough if you were able not to get astonished in front of a billion of matches within a single box.

In my country, the people that do not worry in such situations are usually called lunatics.

That's why you're not able to receive messages from reality.

Should you find all of a sudden that actually you were a disguised CTR, would you get amazed?

Of course, because I'm sure I'm not. It would be a logical absurdity.

Couldn't that be a mysterious message from the cosmos to you?

I'm not a match: my reality can't be that of an information carrier toward myself.

And were you a match? What should be the difference?

I'd be an information carrier, not its target.

Whatever an information may be carried by a billion matches?

It's not the actual figure, it's the diversity: so many matches more than usual.

A very mysterious message.

One must be able to understand the nature of things.

Going on this way, it's very probable that one ends up attributing fancy meanings to random phenomena.

And so, according to you, may it happen that a billion matches fit a matchbox?.

No, it can't

Why, then, are you speaking about random phenomena?

*Should I try to understand why very rare phenomena, but possible one, may take place, for instance an Aurora borealis at **our** latitudes, I'd look for the most plausible explanation, and may be I'd get wrong. Should I try to understand impossible phenomena, then I'd get mad.*

So, because of us, you've got mad.

Quite near to that.

But do you see how the very concept of possibility may get changed?

Life is a heap of successive experiences, through which a congruent reality may be looked for; to the ones who have never met you, you are a pure lunacy, and according to them is insane the one who speaks about extraterrestrials.

And so you've got mad just because you've met us.

May be I'm mad, and what I believe to be true is actually just a creation of my brain; you could not actually exist to other people.

But is your brain able to imagine a matchbox with a billion of matches inside?

(Author: For evident reasons, let's stop here!)

Years ago, I've met, in Novosibirsk, one of the German engineers that had been involved in this story; he was then happily married with Galina, a wonderful Siberian woman, both in her mind and in her body. My German friend has presented me with a transcription of a conversation between his wife and one of the W56's; I enclose it here, without asking a permission from Galina, in this moment rather difficult to find, in the centre of Sibir! The original talk has been in Russian, and I have translated it into English, leaving aside some passages (now, italic will be used for the alien entity):

What do you think of Relativity?

It's an obvious conception, even if Einstein's formulae are but a particular case, a very particular one. Then, Einstein entered a cul-de-sac while looking for the unified field, that actually doesn't exist.

You mean that there's not a single field, from which, according to instances, electrical, magnetic, and so on, fields are generated?

Here Einstein went astray from the very concept of relativity: the way we perceive the nature of a field is strongly relative: according to instances, I may perceive it as an electric one, and you as a gravitational field. One of the absurdities within the Relativity is that absolute qualities are given as granted, far from reality.

But every technology, on out planet, is based upon a few fundamental fields.

That's because you never tried to explore extreme situations, where a field may commute (Author: The Russian verb переключать has a stronger meaning, it means switching from a state to another as in a flip-flop) *from a look to another.*

. . . .

So, you mean that an unified field does exist.

Of course yes, but not in the way Einstein was looking for it: you'll never be able to see a single field, whence electricity, magnetism, and so on come from, simply because such a field cannot be perceived directly. On the contrary, you always see it disguised under different masks, according to the way you are in front of it.

And how may I switch (Author: Again the same verb) from one mask to another one?

You have to get into extreme conditions, put your tensors into a critical situation. If in a tensor matrix the functions within the two main diagonals tend to be identically zero, what happens?

The matrix loses any meaning.

Well, do explore situations very near to this one.

But very high energies should be needed.

Not at all: remember that all you have to do is to take some functions to be almost identically null, not so much is required. In a similar way, you may behave so that the resulting determinant is in every moment almost null.

We enter an indeterminate condition.

Whence you may switch (Author: again the same verb) *from a mask to another one: the second derivatives with respect to qast* [24] *of the value of the determinant will permit to you to decide which direction you'll like to follow.*

Although I haven't presented here the whole text, I believe that just these few lines could allow some open-minded physicist (or mathematician, or engineer) to find a totally new way to look at the reality.

To end this section, I present a conversation held with one of these gentlemen, in the Zanarini coffee-house in Bologna, in May, 1967. Here I appear to explain mathematics to my interlocutor (actually, just from a formal point of view). I was believing that this conversation had remained concealed, but, after having given a copy to a friend of mine, another to another one, and so on and so forth, I've discovered that nowadays it has got to be a public knowledge, therefore I can't but worry any longer. As usual, I'm any way going to cut away some sentences:

Alberto was actually right when he said that he had been eating lamb and fried potatoes together with your people. Who would have expected to see an extraterrestrial drinking a whisky?

What's strange in that? It's worthwhile to accept the positive aspects in the worlds we are visiting.

Is it so long a time that you have been wandering from a place to another?

I'd say so. Any way, one never gets tired to learn new things.

But aren't you homesick? Where are you from?

Well, my home is wherever I am. I do not have a stable home, in the sense you're meaning. I've been born in a place very far away from your Earth. I can't tell you the name of the star my planet is orbiting, because I'm not familiar with your names system. It is a yellow star, a bit larger than your sun. My planet is the fifth one in the system, it is, it too, a bit larger than your Earth. The most important difference consists in that there is more oxygen and less nitrogen in its atmosphere.

And haven't you got any trouble from that?

Very little, and only in the very first days I have spent here. By the way, before we start to stay in a new environment, our biologists do their best to adapt our bodies to it, if differences are not so marked.

How long have you been here?

Some thirty years. At first I have been in China (you wouldn't have expected that, would you?), then in Australia, then in Germany, and at the end in Italy. When I have entered Germany, I have taken a German identity, that I have kept here in Italy. My job, importing German goods, allows me a rather good freedom. Moreover, I remain a German citizen, and I believe I'll keep this status up to the end of my staying here.

When are you going to get away?

I don't know; by sure, within a few years.

Have you stable bases also in China and Australia?

We have been keeping bases on this planet since centuries, I'd say even millennia. Stability is a stupid concept: everything is evolving. At times we shut a base down, other times we open a new one. It depends.

Have you bases also on other planets in our solar system?

A few of them. Actually, they do not make too much sense, because our true bases are some mother ships orbiting inside the system. Only on your Earth, not to be noticed, we have to build concealed facilities, under the ground, under the seas, or in inaccessible areas.

Any way, you are quite free to do whatever you like.

Only if one takes the care to get inserted within your environment, as I have done. Secrecy would present a lot of problems.

Why do you stay in the shadow? Why don't you let yourselves evident to everybody?

First than all, we are not interested at. What could we gain from that? And, moreover, what would be the good for you? In this moment you do not really need spiritual guides, you have to dig your own road by yourselves, because you are able to. Mainly we are here to study, and, by the way, to protect you against possible hostile behaviour from the part of the CTR's. Obviously, in this second activity, we find it convenient to have some support by someone among you.

How do you select Earth persons to get in touch with?

First than all, we look for persons with a good mental stability, gifted with quietness, and the ability of self-control.

Bruno Sammaciccia, therefore.

Yes, but not only him.

And what may you tell me about a phantom "cosmic police"? Someone, I do not remember who, has told me something about that.

It's a structure similar to yours, in some ways, at a higher level, and it is controlled by my companions, at a higher level. It is them that actually control the activity of the ones you call CTR's.

Yeah, all these silly initials. What about W56?

I really do not know. I believe that 56 refers to 1956, although I do not have the faintest idea of anything so important having taken place in that year. About the W, I again haven't the least idea. It is not an initial: were it so, it would refer to the name Weiros, the ones you call CTR's.

What about CTR's?

The same. I can't understand the meaning of these initials, nor do I know who has invented this name.

How many Earth languages are you fluent in?

German, English, Italian, Chinese, Hindi, Russian, Latin, Sanskrit.

Even Latin and Sanskrit?

Sanskrit is a wonderful language, as Latin is.

Swasti Uttara Devadam?

Your pronunciation is awful. You do not know how Sanskrit was actually spoken.

Any way it should have been similar to that.

Yes, I was joking. Any way, if you're interested at, I may try to find for you some recordings from those times; I should before discover who is in charge of them.

Probably, in this moment, I'd be more interested in recordings of classic

Egyptian.

I'll look for them, but I'm not able to promise you anything.

. . . .

Well, your technology is a rather simple one. Once it's clear that any required amount of energy may be obtained through your mother cells, no real problem may be encountered in understanding the way your hardware operates.

Care, things are not as easy as you have been shown. Any way I believe that, had you enough money and enough technicians available, within ten years or so you should be able to build a bell, may be just by trials. Moreover, our bells are really ancient devices, from the point of view of technical design. (Author: A flying saucer was usually named a "bell", or a "scout").

How is it possible that bells cannot be detected by radar?

Because, as usual, your technicians have a restricted view. A radar works fine for the purposes it has been built for, but that doesn't imply that it may be able to detect a bell. When the pilot of one of your planes decides to make a specific manoeuvre, he has only one fixed sequence of operations available. On the other side, the computer driving one of our bells has, in every moment, to choose at random among an almost infinite number of different possibilities, all of them equivalent in front of the required result. That does mean that the distribution of the fields around the bell cannot be forecast, and it changes continuously. Think about what optical lighting does mean. The light from your Sun has an almost flat spectrum, but an object getting light from the sun gives back only some frequencies, and therefore it looks to be colored, and the reflected light has no longer a flat spectrum. In our case, things get more complicated than that, because there is a powerful electromagnetic activity, with relevant exchanges of energy. Moreover, both the emission spectrum and the reflection function are changing in every moment. From a purely theoretical point of view, it may be that these two entities remain roughly constant during a certain period of time. Then, your radar would be able to make a temporal convolution, and so to detect a bell near by. But most of times things do not go that way. There is also a de facto invisibility. During the time a radar needs to look back at the same area in the sky, the bell may have moved far away, and a single echo generates no target. Or, may be, the bell has moved not so far, but two successive echoes cannot be correlated, even if their spectra remain roughly the same. In a certain sense, this is a passive way, that means that the bell does nothing to get invisible. On the contrary, if we decide to do so, we may order our computer to change at random emission and reflection spectra, so that in every moment a closed convolution is simply impossible.

That means that at times you are visible, and at times you are not?

Even worse than that. At times we may be invisible, at times visible, at times visible only in part, at times things may be visible that are not our bell, for instance some of the generated fields. The eye of man, just like a radar, pretends to be an intelligent instrument. It does not convey to the brain an image that appears for one tenth of a second only (or, if you prefer, it's the brain that does not take such an information in any care). When one of your photographers, while developing a film, finds a "flying disc" in a picture, an object that nobody had seen when the picture had been taken, he may say that he has been lucky. The disc was not invisible at all, but the brains of bystanders had not been able to perceive it. The photographic camera is a stupid device, it doesn't try to correlate. If at a certain moment an object is visible, it gets recorded. We do not care so much, because we will no longer be in the area, when the film will be developed. At the end, there is a kind of environmental invisibility. Just like the pilots of your submarine ships are used to, if we stop the resonance, then the bell no longer emits. This way it becomes automatically invisible if it rests within the shadow cone of your planet, on the ground, or in the space.

. . . .

The pilots of your planes have to take care of such things, because their speed and range are limited. To such people a course mistake may mean not to be able to get to the desired airport. And, any way, changes in speed and in course usually reflect increments in the flight time. On the contrary, with our bells, should we even bounce from the right to the left of the course, and back, like the teeth of a comb, we would face no problem in keeping our schedule. By the way, keeping exactly a course is of no particular use: a bell is just a vehicle to get from one place to another one; The way it does so is rather meaningless. Should I be interested to fly over a peculiar point, all I should have to do would consist in placing an anchor over there (Author: An anchor is a kind of VOR, a device that may be detected by the bell's computer, so that it may decide its course in relation to it; from a physical point of view, it may be a small sphere, just 3 millimeters in diameter, but it may also be a "virtual" thing, a kind of "property" attached to a specific point in the space).

. . . .

Very often terrestrials have inquired me about the "Hyperspace", that actually does not exist, because it is an invention from your part. Actually, the space is an entity with a very high number of dimensions, and in every moment we may perceive from two to four of them. Selecting them with care, one may enter into very complex situations. It would be unfair to say that a bell is just a kind of vehicle; situations may take place, where the set of the totality of the points within the universe is such that each one of them has a distance from every other point exactly equal to zero. In such instances, you cannot speak about displacements, and so the bell becomes just a container, not a vehicle (Author: This is General Relativity, stressed up to a point we have not been able to reach by ourselves; from a mathematical point of view, this really extreme idea looks to be sustainable).

What do you mean?

It would be better, were I able to speak correctly your mathematical terms. Think that the whole space sits over Earth surface, with a reference system based upon latitudes and longitudes. If you put up a geometry over this space, everything will work fine until you stay near to the equator, and if the dimensions of your objects will be small with respect to the radius of Earth. But, if you get near to a pole, dimensions along the parallels will be getting smaller and smaller, and points will gather together. In such a situation, you get an universe of zero length. Do you understand?

Sure.

How would you state that, in your mathematical terms?

This happens when one tries to use an Euclidean geometry inside a Riemann space, where conditions are of quasi-Euclidean geometry, and they stand only locally. It's like when applying geodetic equations (the variation of the infinitesimal length is nil), and one finds a simple infinity of solutions, each different from one another.

Fine, what do you mean with variation?

With variation we mean the change of a function, keeping still contour conditions. If I think of all the possible trajectories from one point to another one, on a plane, each one of them may be identified through a function that represents them all. For instance, the function will be a two-variables one. If this function is continuous, its variation is the derivative within the spectrum of the represented functions, with respect to the two independent variables of the representative function.

And so, when the variation is zero, it means that, around a certain value, the functions are identical to each other.

Better to say, that differences between two functions whatever, belonging to a small entourage around it, are of a lower order with respect to the values of the independent variables of the representative function.

And geodesy?

If I put each trajectory in connection with its length, I get a condition of geodesy when the infinitesimal value of its length does not change when the function varies within a sufficiently small entourage. It's a bit like the concept of derivatives, this time expressed as an infinite set of functions, and a single function that represents them all.

Fine, we use a rather similar system, that operates through parametrical representations: a function of functions, at the end.

I'd say that it is the very same thing. Parametry derives from the use of a representative function that, in your words, looks to be just the function of the parameters. Another question: does the word "clacteem" mean anything to you? Williamson was writing that you call your bells with this name.

Not at all. Any way, it's names, may be someone is using the one you've quoted. You too call the same thing with many different names, and may be do not even know all of them, in this moment I'm not able to find an example for that.

May be the hundreds of different names Englishmen call a street.

That's right. If you like, call them Clacteem, or flying saucers. Actually I do not like so much "flying saucers": from one side it's highly reductive, then it makes one thinking of a vegetable soup.

A soup?

Yeah, because of the "saucer".

I do not like too much the term UFO.

Indeed I do, if nothing else, because we have our own UFO's. As one of your writers was used to say, there are much more things than what man may be able to imagine. At times, really seldom, we chance to meet very large objects, in all evidence artificial objects because they do not move within the space in the way natural bodies do. But we have never been able to understand who they are, nor if they are sending messages. Our attempts to communicate with them always failed.

And didn't you try to get near?

It's a paradox: rocket driven vehicles should be necessary, because non inertial devices are repelled.

Why don't you try to follow them?

The Night wanderers (what a poetical name, isn't it?) often change their dimensional environment, so that they escape in an unrecoverable way. May be, one day or another, we'll meet them

Tell me something about the quarrel between you and the CTR's.

You are right in not using the word "war", as Bruno does, because actually there has never been a true war, although many among our friends have died. The name "CTR" has been invented I don't know by whom, in a terrestrial language. These robots have started evolving in a place that I do not know how it is named by your astronomers; it's a star among the ones that constitute the Centaurus constellation.

Proxima Centauri?

No, it is not one of the stars nearest to you; on the contrary, it's rather far away. By the way, your astronomers are not aware that there is a star nearer to you than Proxima Centaury.

Our sun.

How witty are you. Any way, the CTR's are the result of an experiment that has run out of control. They are robots, in the full meaning of this word, even if centuries ago they have started an activity of biological reproduction. To you, at this point, it's no longer possible to discriminate between a natural being and a biological robot.

What do you mean?

Probably you would agree in stating that a synthetized human body, with a conscience and a will impressed from outside upon a pre-existing amorphous structure might be called a robot. But, according to you, its grand-grand-grandsons may still be considered robots?

About us all, are we pre-programmed robots, or aren't we?

Well, according to our biological and – I do not know a correct word in your language – let's say animistic knowledge, it is possible to distinguish natural beings from the descendants from robots; speaking with Bruno, I've found out that a trace exists of this concept within a branch of Yoga, but I do not remember its name.

The Raja Yoga, and, before it, the Bhagavad Gita.

I do not know the names, so I can't confirm you. Any way, it's evident that, in a sense, we all have been originated in the beginning as programmed robots. In a certain sense it's evident that it has been us to program ourselves, because we share God's essence. The CTR's (and, by the way, not them only) are on the contrary artificial creatures, and they remain as such, even if nobody among your physicians would be able to discriminate; actually, there is nothing physical that may allow to state whether an organism is a descendant from artificial robots, or not. The CTR's, also named Weiros, are therefore an artificial race, and their main goal consists in trying to fill up the gap between them and natural races, and therefore they are studying. As far as I know, up to now they have only been able to ascertain that the problem exists, but have not been able to understand it, in its full details.

And why do they fight against you?

Indeed, we do not fight, we do not consider this situation as a war, because no war may issue between natural creatures.

How fine if some of our top brass were able to listen to you.

Any way, getting back to the CTR's, their mission consists in trying to solve a problem, whose proposition they do not know in detail. To them it's a matter of vital importance, and therefore they do not surrender to any difficulty. Nowadays they have got an extremely high culture in biology, much superior to your own; hopefully, we remain in advance, at least because it has been us to create the CTR's.

What do you mean?

The experiment that has given birth to these robots has been made by a natural man, who, unfortunately, has died while trying to stop the process. Therefore the CTR's are "sons of man", as your Bible states.

And man is son of God.

Not at all, man is God. Man is part of God, who is not even in his smallest part foreign to man. There is no feature of God that is alien to human nature, at least not at a physical level.

The Raja Yoga again.

Yes, it, or something like that. To schematize, let's remain with God, men, and CTR's. The biologic know-how of the CTR's would be astonishing to you; just think of Hydras (Author: We'll be speaking later on about this "hot" subject), *and you'll realize that, in a certain sense, their biology goes beyond physical death, the point where your medical knowledge stops. But they are not able to understand what they're missing in order to fill the final gap. And to them the only way to try to understand consists in studying. And men are the object of their research. In practice they try to experiment on the men who are so weak to be unable to defend themselves, or not even to realize that they are in danger, you for instance. The men who are able to realize that such a problem exists, we and our friends, try to prevent other men from being damaged. That's what the so-called war consists of; actually it's like when you fill the air around you with vaporized poisons to kill insects. Think of insects with technological capabilities well beyond yours, and you'll get an idea of the situation. Hopefully, our culture is greatly more advanced that theirs, therefore we are able to stand foremost. But, at our levels, even a small gap results into something really huge, compared to your situation, and any way the weakest party is extremely more powerful than you.*
Our forefathers had already got in touch with yours; many of your religious teachings have been originated by us. The fight between good and evil is in the way your friars are depicting it, not as your priests say. It is not bad, in itself, to kill a man. It becomes a sin if this action consists in a deviation from the theoretic humanity level, and in indulging towards the CTR's

way of thinking. The evil your priests speak about actually does not exist, because it is foreign to human nature; only when a man tries to give up, even only in part, his nature, then he acquires part of the features of the CTR's and of people like them, and that is evil; the dualism exists between who is human and who is not; you have named good the first and evil the second one, and that's all.

Yes, that's rather well what the texts I mention maintain. Any way, a couple of times you have hinted that there are not only the CTR's.

Quite true. Or, if you prefer, we must agree on names. If with CTR's we mean the people we are struggling with in this moment, the ones who have killed the daughter of your friend, fine, they are not alone. If we refer to the whole of non-natural races, well, in a way or another, we may name CTR's them all.

And what about natural races?

There are many kinds of them, although the human one is by far the most frequent. There are, for instance, the Wan people (Author: Beings whose flesh is almost transparent) *who, as far as we know, have been this way since ever. Then, there are races totally different, from a physiologic point of view. You would be surprised to know that on your Earth there's a race of beings you've never noticed.*

Who are they?

You maintain that curiosity is feminine.

Well, you'll understand that one gets a bit curious, at this point.

Latin is one of the languages of yours that I like most. There's a Latin sentence that could represent a very good answer, but now I do not remember it. Aside joking, these two races have lived in the same place for so long a time, without interfering with one another, probably because it would be very difficult for you to get in touch with them. There's no reason to alter this situation, it could even be dangerous.

Are they aware of us?

Yes, but they do nothing. Don't care, it was just some kind of a joke from my part. As there isn't any interference, you may think that they do not exist.

Roger, I'm able to recognize when it's useless to go on. Then?

Then, there are races, still human ones, that live in dimensional levels different from yours, and so, once again, to you they do not exist.

May you propose me an example of a non-human race?

The monsters circus, isn't it? You'd be surprised, because non-human implies a totally non-comparable physiology. Think of a rock: could you classify it as a race?

Yes, because I'm able to classify it. I'm able to state that an object is a rock, and another one isn't.

But I'm speaking of races, not of names. A rock may be an individual within a race if it is aware of itself, if it has a will. By themselves, rocks are not a race, although races exist, whose individuals are rocks, from every point of view.

They get born and die?

To achieve an individuality, an internal cohesion is required, one based upon loneliness, or a plurality. If you cut a nail of yours, you afford a loss in your individuality that you may rightly neglect. Were your nervous system be cut in halves, even if your organism were able to survive, you'd be destroyed as a reasoning entity. That's the very same with these pebbles.

Who are not the non-human race you were speaking of before.

No, quite a different thing. What I'm naming pebbles are to be found in the open space, near to a big red star.

Aldebaran in the Taurus constellation?

I don't know. In which direction the star you named is lying?

I don't know. We ought to look into a stellar atlas.

Well, any way I can't see why you care. If I remember rightly, it should be a star some 200 light years away, some fifteen degrees above your ecliptic.

I do not remember how far Aldebaran is, but the direction should be right.

You're always incoherent: you have tired me (Author: Here my interlocutor uses a very coarse Italian word) *speaking about Relativity, then you do not conform to your habits. What could you care with pebbles being near Aldebaran, or anywhere else?*

That's right. Just a bit of curiosity. Changing subjects, what are silicon webs?

What?

They have named them silicon webs: sometimes, after a saucer has flown over, from the sky long filaments slowly fall down; they are some metres long, and they tend to sublime in a rather short time.

What do they do?

They change from a solid state to a gaseous one, without getting liquid in between, and so they disappear.

I don't have the slightest idea, it's a totally new phenomenon to me. May be the fields the bell generates may gather together molecules from the air. Why do you say silicon?

Because, analyzing this substance, it has been found to be composed by boron, silicon, magnesium, a fourth element that now I do not remember, plus an organic volatile component.

By organic you mean long molecules?

Yes, in a certain sense. In chemistry we call organic carbon-based molecules, that are typical of living organisms.

And you don't even know what life is. I repeat that I haven't the list idea about it, but if the substances are what you say, they must come from the atmosphere: our bells do not go around spreading magnesium and silicon. Unless...

Unless what?

Have you any idea about the organic matter?

No. As far as I know, nobody has been quick enough to analyze it.

As usual, probably you haven't understood anything. You tend to frame everything within your mental schemes, without asking yourselves whether there might be a different answer, outside them. I can't be sure that I'm going to give you the right answer, but by sure you would never have thought this way.

Go on.

Any way, it's possible that I'm going to say silly things, because I do not know the details. One of the external groups coming to your Earth is very interested at insects.

Insects?

Actually so. Insects are really rare within the universe. Usually, when they are found, this means that the planet is a young one, because insects are among the first races to appear, and among the first ones to disappear, I don't know why.

Their structure is rather rudimental, they have serious problems in breathing.

Well. On Earth it's different. I'm not sure, because this is not my field of interest, but yours is the only mother-planet to be breeding an insect population. It may be that the silicon you speak about is the by-product of an operation of an environmental analysis conducted by these people. Magnesium, as usual you aren't aware of that, is a powerful catalyst, in insects populations.

What do you mean?

It strikes heavily, both their behaviour and their actual survival. It's the best destroyer of insects. Within a few generations, insects develop a mutation that destroys them.

Something like our cancer.

Not really so, at least because you cancer is induced by yourselves, with environmental contamination, while in this case the mutation gets induced from outside. Any way, the final result is the destruction of their race. It is also possible to make measurements saturating environment with magnesium, in a way I do not know. Any way, it's just an hypothesis, and I can't tell you any more. It's strange that it's a phenomenon I've never heard of.

It is not so frequent.

Then, probably, I'm right.

Let's change subject once again. What is your life? What do you do, what are your aspirations, which problems are you facing?

Well, I may understand your interest, and will try to answer your questions, if then you are going to answer the same questions. To us science is not an important goal. We believe that what we already know is enough by large, therefore our scientists have a secondary role in our society. In a certain way, we are like the CTR's, willing to make a jump in our existence. We still have too many links to the physical world, much more than what we'd like. That's true with ourselves: other people are beyond us, and other behind, you among them. We are sure that man is not so much made of matter, therefore we are trying to get rid of this useless appendix, encouraged by the fact that others have already succeeded. Man's ultimate goal is God, and that looks like the serpent in you sagas, who is biting its own tail.

Ourubus, or something like that.

It's not the name to have importance. Obviously, while tending to God man finds himself. And that's what we're trying to do. You may see that we and the CTR's have similar problems, but there's a big difference: we do not have at our disposal a herd to use in our experiments. Therefore we try to understand how human societies are organized, looking for a common sub-structure, that actually is still escaping our efforts. For instance, we haven't yet understood how it may be that you, a relatively primitive people, exhibit parapsychical attitudes that we need dedicated hardware to grasp. We haven't understood where is the difference, if there's any, and that's one of the subjects we're studying. In this moment, I'm studying you much more that your studying me.

I'm well aware of that. Any way, I trust you, so there is no problem.

Now it's your turn to answer your own question.

From my own point of view, or from the whole of earthlings?

Do you feel confident to speak in the name of you all?

Of course I don't, so I'll speak in my name only. I believe that two significant moments may be identified: looking for a satisfying knowledge and life style, and to get to be able to reach them. Let me explain. I believe that a reasonable way of life is near to what you've been saying, and that I name Bhagavad Gita. But of course, to be able to behave that way, to us it is necessary to have found a job that allows us to have a sufficient free time, and allows a sufficiently high cultural environment. I've studied dead tongues, religions, philosophy, in order to try to understand something. After all, the point I have reached allows me to consider this talk just as a mean to satisfy an years-long curiosity, but do not believe that I'll get significant operational results from it.

I'd be rather surprised of the contrary. From a petty point of view, it's fine to be looked at as if gods, but, after all, such a talk in front of a whisky glass is much more constructive.

To me the problem consists in that our society is not oriented this way. My aims consist in acquiring culture, first than all from within myself, then at a technological level. But technology is a secondary aspect. If my environment would allow it, I'd be an ascetic, on the top of a mountain.

. . . .

Your society is based upon money, and this forces to undesirable compromises.

There have been attempts to overcome this problem, or at least to reduce its implications. Unfortunately, they didn't look to work.

That's because you are tied to individual property. I've found that people here think I'm lucky because I own an aeroplane, that's beyond what most of them may hope. What should happen if I'd leave the plane on the runway, available to anybody?

First than all, if a child gets into it, takes off, and kills himself, it would be your fault, both under our laws and according to my opinion. Then, as it is not a so common object, it is necessary to put up a system to manage the attempts to access it. In our case, such a mechanism is money.

To us there's no need for it, because we may have plenty of them. But societies like yours tend obviously to exalt money for itself, instead of considering it just as an access manager.

That's true, it has always been so during the whole history of our civilization. Since a century ago other systems have been proposed, i.e. communism and the like, but in practice they wouldn't work. But it means that there is an unconscious need to change things.

Would you like to enter our federation?

Whom are you addressing? From my own part, probably so, but I'd need to reflect. All earthlings? I don't think so, probably mainly because of fear. Were they to know that a war exists, or, if you prefer, a stormy condition exists between those strange objects that fly over our heads, probably they'd like to listen to both parties, in order to choose the best offering.

Are you so materialist?

Probably in this moment I'm more pessimistic than usual, but I fear that things go this way on our planet. From your side, who is in charge to take decisions?

I asked Bruno, and he told me that our system is called an oligarchic one: a group of people decide in the name of everybody, and everybody decide who is to decide.

Again the serpent biting its tail.

Not at all. You would be right in a selfish environment. With us, the single person is worth nothing, and is aware of that.

Well, may be you're right. Any way, it looks that we must stop here, because the cafeteria is closing.

What are you going to do with your recording?

Nothing peculiar. I'll translate it into Russian as usual, then I'll file it.

I don't understand to what purpose.

None at all. As I'm used to say, it's a collection of stamps.

Some Further Detail

In what follows, making also use of the notes by Hans, I'm going to present some concepts that are typical of Amicizia (in many instances, it's the first time that things like these are printed); as usual, the list is far from being exhaustive: I'm going to present only concepts that have been referred to in the previous pages.

Bases

It has been told widely that our friends had built a huge underground base, very deep (in practice, lying on the bottom of the continental clod), and many more others, typically small ones, at inferior depths. The big base was a reference point for all of their European activities, and was typically hosting machinery. The smaller ones were used as living places. They were saying that they had some other bases within our solar system, but that they weren't stable structures. Actually generating (and destroying) a base was a very simple process to them. Let's think of two linear magnets, initially parallel to one another; if I start rotating one of them around its barycentre, at first the force lines will get more and more inter-twisted, as the angle between the magnets increases; after a critical angle, the lines of force of the magnetic field will change abruptly, arranging themselves into a new disposition, that prevents them for twisting too much. Our friends on the contrary were able to generate what they were calling a "magnetic tress", i. e. a structure where the lines of force were strictly twisted the ones around the others. Such a thing had the property of "opening" matter, that it to compress it sideways, squashing it on itself; translucent walls were resulting, almost crystal ones, with an astronomical density, a Young module equally very high, and an unbelievable strength. This way they were able to open the cavities that were to become their bases, evidently without damaging in any way the tectonic structures around, on the contrary probably strengthening them. Such a structure remained stable until the fields that had generated it were active: it was sufficient to switch these fields off, a finger over a switch, to get back at once to the *status quo ante*. In a similar way they were opening passages to access their bases, when they were needed, closing them immediately when no longer in use. Only very rarely (very small bases just under the ground) stable corridors were used. Large bases were also big airships, typically orbiting our sun on planes very tilted with respect to the Ecliptic.

The "Hydrae"

Bruno tells of the encounter between some ten earthlings and a strange entity, depicted as a living tar wall, West of Montesilvano.

It must be said that the W56 conceptions about human entity were very similar to the one professed by old Egyptians: according to both, man is made of an individuality (the Egyptian <u>ka</u>), an interface to the third entity (<u>khw</u>), then this last one (<u>bah</u>), and finally the physical body. The <u>bah</u> (to use its Egyptian name) is the entity that drives the body, and is characterized by the most brutal animality. It's where instincts are located, and, in certain sense, it's the entity that drives wild beasts.

Typically, when a man passes away, these three entities get free; within a few hours, <u>bah</u> and <u>khw</u> die too, while the <u>kha</u> goes its own way. Seldom, the <u>bah</u> survives the death of its body, and in such instances it is the cause of the (very few) cases of devilish infestations that have not been debunked by a serious scientific analysis.

The W56 and CTR biological capabilities were astonishing; in particular the CTR were used to seize the <u>bah</u>'s that were temporarily free, after their physical body had passes away; when they had put together some 15 of them, they were able to "melt" them together, using a peculiar device, giving this way birth to a new entity, that the W56's had named "Hydra". It was an entity characterized by pure animality, at the worst thinkable level, aggressive and violent toward every living entity (plants, animals, men). For some obscure reason, the W56's had told that they were able to keep them under control, but not to be able to destroy them. For such an operation, some 10 earthlings were required (again our supposed para-psychological superiority), very sang-froid people, better if used to yoga techniques, plus a long (and heavy) copper chain; the operation was consisting in reaching the Hydra (often in inaccessible places), surrounding it and, clenching at a same time the chain, exerting the will to annihilate that shame. Not always the first try was successful. When the goal was achieved, the Hydra (usually a greenish sphere, a few metres in diameter, differently from what Bruno had lived) exploded, uttering a terrifying shout (it's difficult to understand which phonation organs could have been the cause of that), and its remain were falling on the ground, that slowly was starting to absorb them.

Looking for bah's is one of the reasons why UFO's are so often spotted over cemeteries, around hospitals, on battle-fields, that is in those places where it's easy to find recently died people.

For totally different reasons, the CTR's are also frequenting places where pathos reaches high levels: stadiums, concerts by famous rockstars, political gatherings.

The "aniae"

Under the name "ania" (I invented the English plural "aniae"!) two different set of entities were intended (who knows why, the same name for different objects).

In the first case, an ania is what Williamson refers to as a "black crystal": a very black object, to the point that it looks to absorb the light, an irregular polyhedron with a diameter around one cm. In Italy we were speaking of "valvoline" (that is, little tubes, little valves), today they would speak of "implants", although the technology is quite different. Such objects were inserted inside the body of people, both to their knowledge or not; in a very short time, the ania was disintegrating itself into a myriad of microscopic biological robots, each one of them migrating to the body area it was to work in. The goal of such objects could go from a simple increase in sensitiveness, to the development of telepathic capabilities, up to transform the man into some kind of a slave.

The fact that an ania was starting its job disintegrating itself into many microscopic objects makes me suspect about to-day "implants": weren't they natural objects (that our medical science considers, although classifying them as extremely rare), they would reflect a rudimentary technology in front of the one exhibited by the CTR's and the W56's.

In the second case, an ania is a small biological robot, some 1.5 cm long, some 0.5 cm in diameter (but they may be even smaller); in this case, typically the plural was used: aniae. These small robots were demanded to monitor their environment. If it is required, they are able to grow wings and small paws, so that they may be taken for blow-flies. Their secondary activity consists in ensuring their own safety. While alone, these objects aren't too dangerous; on the contrary, when in group, they may be extremely harmful: for instance, they may extend themselves between two wires of electrical power, to cause a short circuit; they may clog the Pitot tube of a flying airplane, or the oxygen supplier of a scuba diver, or they are able to saw the steering rods of a running car (all such exploits have actually taken place), or, even worse that that, they may act as a beaming homer to concentrate into a small area violent energies generated elsewhere.

An interesting detail: when a single ania activates itself, typically it gets lit for some seconds; when a group of aniae activates, a chirping noise is heard, as from cicadae.

Dead people

Casualties have been recorded in both sides, during the quarrel between the W56's and the CTR's (they were refusing to speak about a war), both among themselves and also among the earthlings that were supporting one party or the other. For instance, a small girl, the daughter of a German couple, who were in the W56 area, or a lady, sentimentally tied to the chief of the Italian W56 group (she had even got a son from him!), the unaware passengers of a DC3 that crashed near Ciampino airport (at the time, they were pretending that the aircraft had been forced into the ground, with the aim of killing one of the W56's, who was on board) [25]. The war – or quarrel, if you like – has therefore a rather cruel one.

Then a case, which evidently ran out of control: a summer night, in the 60's, all of a sudden tens of injured men appeared in the streets of Pescara; they were taken to the hospital, they refused any assistance, and disappeared during the ensuing night! Local newspapers have discussed this strange incident.

Satellite disks

For reasons inherent to its working principles, a scout faces serious problems in acquiring information about its immediate environments; this problem is obviously the more serious the lower the scout's height is. Because of that, typically while at low altitude, a scout emits some small flying robots (some 10 cm large), who start flying around it, transmitting information about the whereabouts; without their help, low flying would be extremely dangerous.

Looking carefully to a picture of a low flying scout, often such satellite discs may be found (look, for instance, at the picture no. 31, in the Appendix); actually such robots usually are not disk-like, but this is the name they've got.

The Карманные Скауты

In all evidence, a Russian name, whose pronunciation is, roughly, "Karmanneeaiee Skoutee", from the German Einsteckbar Skout; at those times, I was used to write down my notes in cursive Russian, because my handwriting is so ugly that Russians themselves would have been unable to read them! The name means "Pocket scouts", and refers to objects whose purpose I've been unable to understand.

I'm speaking about small scouts, some 9 metres in diameter, that, in some way, were compressed to a diameter of some 40 cm, with a similar reduction in their mass and inertia (actually an inertia sensibly higher than one would expect, in relation to the mass). A Карманный Скаут, when compressed, had a weight of just some three to four kilograms. Any engineer will recognize that, when compressed, a Карманный Скаут was an extremely delicate, even brittle object (I believe; I have seen one, but just touched it with the tips of my fingers), because its mechanical resistances are reduced with the second (may be even the third) power of its linear dimensions. It was kept inside a square rigid bag, some 60 cm wide, without an handle, just to make things easier,

Having got to a rather wide clearing, the small scout was taken out from its bag, and put down with care; then, one had to get at least twenty metres away, if possible concealing himself behind a tree or a wall; acting on a switch inside the bag, the scout would at once get back to its original dimensions (with an obvious violent blast, pebbles shot like bullets, followed by an inverse air displacement, a strong sound, and leaves flying wildly everywhere; in a short time things were quieting. The scout was now ready to be flown. When everything was over, the inverse operation (same blasts and noises); this second operation typically was generating lower gradients of pressure, therefore it was not as violent as the first one. In both instances, sensible variations in the air temperature were felt.

I have never understood the use of such devices: it would have been much easier to have a scout, on auto-pilot, following its owner at high height, and have it land when it was necessary. From time to time the technology of the W56's was presenting obscure aspects. Probably there was a reason behind such complicated devices, but no satisfactory explanation has been given.

The Карманные Скаути were the discs given to the (few) earthlings who had learned how to master them. Probably the only purpose of such objects was in relieving their owners from the problem of concealing somewhere a 9 metres scout, but even so the complexity of the process was not so easy: going to a secluded spot, "inflating" the scout, then flying away leaving one's car in the spot, and so on.

Overalls

A bit less rare among earthlings than the Карманные Скауты, overalls were biological entities, strictly personal; they acted at a time as defensive systems and as transportation devices. They were to be worn over the naked body, and they were protecting their occupant against practically whatever external danger.

There was a whimsical system for managing inertia: from one side, the overall was able to increment its inertia up to unbelievable levels: they were maintaining that a guy, sitting over an atomic bomb, after the blast would have found himself at that very place. Then, the propulsive mechanism was a really rudimentary one, based upon two pushes applied perpendicularly to the soles, and the only control system was a button, on the belt, that was able to modulate the intensity of the two pushes (identical to one another). Pushes could be both positive (upwards) and negative (the other way). The pilot's skill was in graduating the strength of pushes, and in orienting carefully his feet. Useless to say, more than once funny episodes took place: for instance, one morning in München, in Germany, astonished passers-by have been looking, perplexed, at an elderly distinguished gentleman who, upside down, was flying randomly at an height of a few meters, knocking from time to time unto the buildings on his path!

The overall would tune itself to his owner identity, that it was able to recognize not through the DNA, but thanks to a biologic principle still unknown to our scientists. A different guest would have been considered a potential enemy, and therefore it should have been destroyed.

Some typical jokes

Being in touch with such people involved a deep friendliness, with frequent actions, from their part, that I might just qualify as "jokes". Here I present some of them.

Street lamps

It was very frequent (at times even to-day) that if one of the group was walking along a road lighted by mercury lamps, a "shadow zone" was generated, that moved with him; that is, the lamp he is approaching switches off, to switch on again when he has got beyond, then the next lamp switches off, and so on. Such a phenomenon happens both when walking, or when driving a car. Paolo Di Girolamo, in his future book, is going to relate on many cases of this kind.

Mad little machines

From Bruno's tales it results evident that the activity of supporting our friends was an extremely costly one: two trucks full of fruit twice a month, just to quote one expense; Bruno was taking charge of most of expenses, others were contributing when they were able, and willing to. At times our friends were trying to help, in a way or another (Bruno tells of a fall of platinum ingots into his garden!), and sometimes they produced whimsical devices. In a couple of occasions they had given Giancarlo small machines that were to produce valuable items, that one might then try to sell, or things like that. In both instances, the devices resulted totally useless, because what they were producing had no commercial value at all (it would be too long to explain why). In return they were absorbing electrical power as sponges do with water, taking it from power cables near by, and therefore resulting in (obviously) astronomic bills!

Where They Come From

Obviously I am the first one to state that an information that can't be verified is void of any meaning. Any way, just for the fun of it, I present here part of the list of the stars they were pretending to be coming from. It may be interesting to observe that, according to our astronomers, almost all of these stars should not be able to host life, as we know it. But, since the Cepheid variables, our astronomy seems to indulge in indefensible activities, so I present this table any way.

Of course, I'm unable to tell whether it is true or not, nor do I believe that anyone will be able to show that it is right, or false. I am convinced that it is plausible, although "officialdom" would reject it as a matter of principle.

Name	Name	Spectral class	Absolute luminosity	Distance (l.y.)	General characteristics
α *Aurigae*	*Capella*	G0	-0.5	42.2	Quadruple giant
α *Bootis*	*Arcturus*	K0	-0.1	36.7	Giant
α *Cassiopeiae*	*Shedir*	K0	-2.0	228.6	Giant
α *Tauri*	*Aldebaran*	K5	-0.6	65.1	Giant
β *Andromedae*	*Mirach*	M0	-1.9	199.4	Giant
β *Cygni*	*Albireo*	K0	-2.0	385.5	Double
β *Geminorum*	*Pollux*	K0	1.1	33.7	Giant
β *Ursae Minoris*	*Kochab*	K5	-0.9	126.5	Giant
δ *Draconis*	*Altais*	K0	0.6	18.8	Giant
ε *Aurigae*	*Almaaz*	A8	-6.1	2038	Double super giant
ε *Draconis*	*Tyl*	K0	0.8	102.2	Double
ε *Tauri*	*Ain*	K0	3.5	155	Double giant
η *Tauri*	*Alcyone*	B7	-2.4	368	Double
ι *Draconis*	*Edasich*	K0	0.8	102.2	Giant (+ a planet)
19 *Tauri*	*Taigete*	B7	4.4	372	

It must be noted that our friends were not pretending to be the only aliens wandering on our continent; at least two more groups were present: the UTI (again a name invented by Bruno; I do not know its meaning) and the Elta V (again I don't know the meaning of this name; in this case it had not been invented by Bruno). The CTR's and the Elta V where coming from these stars:

Name	Name	Spectral Class	Absolute luminosity	Distance (l.y.)	General characteristics
δ Pavonis		G8	4.6	19.9	Yellow dwarf
ε Orionis	Alnilam	B0	-6.4	1,340	Supergiant
ε Ursae Minoris		G5	4.2	200	Triple Giant
ζ Orionis	Alnitak	B0	2.0	1,500	Triple blue Supergiant
τ Ceti		G8	5.7	11.9	
61 Cygni		K5	7.1	11.4	Double

Of course what I've said about the first table is worth also for this second one. It looks that almost nobody, among our visitors, has been born under a star similar to our sun!

Maieutics

Although I have not been deep inside this saga, I can't but recall that our friends have given us a quantity of information, almost on every branch of knowledge, in peculiar on physics and technology (speaking of engineering should be both reductive and misleading).

But I believe that their, let's say, "moral" information are by far more important. It is not so strange that an engineer appraises subjects orthogonal to scientific topics, but years of practice of yoga (both from a physical point of view, and from philosophic aspects) have made me perceive, since when I was a boy, that there is a whole world worth to be explored in these directions.

For what strange a reason almost all who got in touch with this story have named it "Friendship"? Because in the behaviour of our friends friendship was their motive impulse. They were used to say "I am everybody, and everybody are me", a concept handed down only in part from the gospel tales, that have been often corrupted for not so high purposes. Are we able to realize what this small sentence means, what deep implications it reveals? Jung has got a bit near to it, with his collective unconscious, nevertheless a vision that limits reality; Vivekananda got near to it, in his commentaries to the Mahabharata, but he too had only a partial vision. That small sentence enlightens a concept that we feel within our heart, but refuse to understand: all living entities, from microbes to luminaries, are the very same thing, because they reflect, in a way or another, a single reality, God. Even more: also the so-called inanimate entities are indeed participating in this reality (who had thought such a concept to the unknown authors of the yoga philosophy?). Therefore the dualism between the "myself" and the "outside myself" is meaningless. Dear friends (this name came involuntary) we are all the same thing, I who like to discourse on fractal analysis, the madman who explodes himself among people, he who rapes boys, mother Teresa from Calcutta, we are one and the same, that entity that the W56's hade named *Uredda* (actually this name refers to something near to it).

The so-called individuality (Kant, Descartes, and so on) derives from a mistake: usually we believe to be what we think ("Cogito, ergo sum"), we believe to coincide with our sentient mind. It's really wrong: mind is but an instrument, not our essence, that is God, as told in short by the Sanskrit sentence "Tat tvam asi". Identifying ourselves with our mind, or, even worse, with our senses, leads us to the variety of attitudes, from the rapist to the terrorist, to the saint woman from Calcutta.

Yoga philosophy (and, to read accurately, also the Gospels) damns as baseness retiring from the world (one should read that monument of moral science that is the Bhagavad Gita), and actually friars, the ones Bruno was loving so much, do not shut themselves up into an ivory tower; on the contrary, while trying to overcome mind-induced individualism, they live together with humanity, trying to be of some help. Brother Francesco was used to torment his body, subjecting it to the worst insults, in order to resist the ill-omened influence by his mind.

The W56's had mastered this concept since aeons, indeed they were living at ease with their bodies: they were indulging in the material pleasures of life, without letting themselves to get influenced (we are still fully in the Raja Yoga): they were using their body and their mind just like instruments, things to guard and to be attended to, but never to be seen as the quintessence of their entity, of the human, and therefore divine, nature. I don't know whether Brother Francesco, at his times, had got in touch with similar beings, but he has perfectly staged their morality, and it has not been by chance that Bruno has always admired the Italian friar. Were I to name the book that best describes the moral of W56's, I'd suggest Francesco's *opera omnia*, may be the version sponsored by Bruno himself, in an effort to spread such concepts.

Indeed, it's even too easy to demonstrate this truth: were we coincident with our mind and our body, what would happen, for instance, when we get drunk, and both our body and our mind are temporarily out of order? In those moment are we outside the cosmic context? What happens when a barber cuts us a lock? Although minimal, it should always be a loss of individuality (see "Metamorphoseon", by

Apuleius from Madaura). Actually the magicians of his time weren't totally wrong, as also the W56's were maintaining, and the precautions taken by the barber in the novel underline this fact [26]; but in this case it's quite a different subject, it refers to physical integrity, to the khw, not to the ka. How it happens that after one dies, his hair, beard, nails, continue to grow for a while? Is it still his individuality to manage this phenomenon, or what?

Evil: a very hard concept. One of the W56's had once told me that it is not fundamentally evil to kill a man (within certain contexts); were our civilization able to receive such an idea, first than all if should get rid of laws and judges (that actually do not exist with the W56's – one can't find what could be their use). But, from their point of view, evil is a really immanent concept, a challenge to everybody against himself. "Evil – they used to say – is a deflexion from human nature." Men, as every living beings, are characterized by what we engineers name an envelope in the space of phases, in short by certain kinds of behaviour. The error that comes from a justice administrated by men consists in that a certain behaviour is declared wrong in theory, and is punished, without getting any positive result this way, on the contrary often worsening the situation. The day we'll be able to cancel this useless appendix of the state, we'll have made a first, short, step towards civilization.

If someone wrongs me, and I resent it, I am guilty any way, because I apply my petty apparent individuality to a phenomenon that reserves no importance on a cosmic scale. That's why Someone said: "Turn the other cheek." On the contrary, if I get a success, and so I am a winner, I must immediately win once again, this time against myself, in order not to get ride from what has happened. The name Bruno had given to our friends, W, comes strictly from this concept.

State: what's it? An entity that qualifies itself (both in dictatorship and in democracy) as the emission of the sovereign people, that in the name of its subjects emanates laws (see before), applies taxes, and stands as a parasite behind the so-called sovereign people. If we think a moment on that, we may realize that it has no use at all. Even more, were all the nations to disappear from our planet, together with politicians, of whatever tendency, earthlings could believe to have made another little step forward. The W56's had no idea of a state; someone, selected on an oligarchic principles, were dictating general rules, and that was all. When I asked whether the selection was made, say, by a computer, in order to warrant equity, my friends burst into laugh: "And why? Some among us could be interested to influence the election with malice?". The W56's are really light years ahead of us.

Science. As a technician, I am obviously fond of science. But science too, to the W56's, is only an instrument, not a final goal. I may say that I've learnt more from some of my pet animals than from high level scientists. Our friends were of course on a privileged level, because of their unthinkable technology, but they were maintaining an attitude coherent to themselves, they were not considering science as particularly important, they were pretending to know already more than enough, and so they were making researches, but almost for the fun of it.

God, Allah, YHWE, and so on. Also on this subject the W56's were following the classical yoga philosophy, the "Tat tvam asi" (= "You are your god"). Man is God, and the divine entity coincides with the set of all living, and all non-living entities. Care, it's not a trivial pantheism, that doesn't mean that there's a god concealed inside the pebble out of my door. It means that nothing (also at levels still unknown to us) is in any way foreign to the nature of God. I repeat, I, you, Mother Teresa, the children rapist, the terrorist, we are all God. Let me use a technicality (I'm not a theologian!), God is but the envelope (in a mathematical sense) of the characteristics of whatever entity. Earth religions have got near to this concept, with different result, but unfortunately the mirage of temporal power have always bent them towards more trivial attitudes. Although they were respecting whatever creed of our planet, the W56's were maintaining that, according to them, there is no need for rituals, of worships, of asking for grace: God is within us, we are God, eventually we should ask for a grace to ourselves (it's even much more true than how I've been able to explain here).

While Bruno was dictating me his memories, in some occasion I tried to force him to afford such subjects, but he had refused to discuss them plainly. "People – he was used to answer me – do not want to understand certain things, even when they are confronted with them.". Some of these concepts are well examined in the book by prof. Marhaba, quoted in the bibliography, but unfortunately it's in Italian, and I fear that now it's out of print. Of course I'm not able to offer the reader a talk with Dimpietro (I myself has seen him just once, by chance, and I regret not to have stopped him in order to spend some time in discussions). Nor may I invite the reader to a tourist trip inside the nearest base of the W56's (as Bruno was accustomed to). Any way, I believe that it would be difficult not to share these last sentences, although he who has written them is an earthling, and not an alien from aboard his flying saucer, an earthling who has been believing such concepts since ever, being only later supported in his opinions by the W56's. These latter were almost never speaking *ex cathedra*, on the contrary, during our talks, they were exerting a kind of maieutics, extracting from within ourselves concepts that we had been kept buried under tons of commonplaces and convenient attitudes. By sure, this has been the most important gift by Amicizia, although at the end it has been just the Earth side to have failed.

When I was discussing with Hans of Generalized Relativity (which I am fond of), he made me realize that the actual conceiver of Relativity had postulated absolute entities: space, time, and all of the involved fields (please refer to the talk with Galina, the Siberian physicist). Why an electric field must be such, and keep its identity in a static way? Maieutics, as usual. Before I met the W56's, I had endeavoured a re-writing of Generalized Relativity, just in this direction. An incredulous reader might suspect that I have invented aliens to support my own ideas. Were it so, what changes?

On the contrary, let's take a pencil and (many!) sheets of paper, and re-write Generalized Relativity in these terms: space, time, fields, are all complex entities (in the mathematical sense of the term), that may be described using quaternion algebra. He who understands a bit of complex analysis will have, may be, a small illumination: the number

4... If I apply all that to tensor analysis, I may find and infinity (not in the mathematical sense) of possible different situations. I ask the reader without a specific culture on these subject to believe that this is the only possible way out in the panorama of contemporary physics, bogged down in strange discussions about sub-particles that have not an existence of their own, because they are generated during experiments; as usual, it is useless to discuss whether these ideas are of my own, or come from outside).

Many concepts that physics presents as absolute and fundamental ones, are actually mere appearance. Not so long ago, discussing with a friend of mine, I've shown him that the absolute zero, and the second principle of thermodynamics may be discussed at length (years ago I have lectured at the Physics Faculty in the University of Bologna, discussing the same concepts). Nothing may be taken for granted, not even axioms. One of the messages the W56's have taken from us is that everything must be discussed (of course, in an intelligent way). There is a funny theorem of plane geometry (see Appendix) that asserts that all triangles are equilateral ones, and it is usually presented as a paradox. On the contrary, the theorem is correct, and the mistake must be looked for inside Euclid's "Elements". It has been Euclid to go wrong, because he had never formalized the concepts of "internal" and "external", therefore, according to his "Elements" this theorem is right. Yet, since centuries, we have been presenting Euclid's geometry as an example of logical coherence. It is true that since some decades they started discussing the meaning of axioms, but, unfortunately, the discussions are limited to a few mathematicians.

Since years I've been going around lecturing against the idea that the speed of light in a vacuum is an upper limit to speeds, and actually speaking in relativistic terms, starting from the obvious fact that Einstein himself had never dared to state such a limit. It is true that we are not able to speed an electron inside a vacuum tube to more than a small fraction of the light speed, but as usual we are mixing causes and effects.

The relativistic c is, at all effects, and for what concerns us, the speed of light in a vacuum, but Einstein himself had stated that it was "… an experience-based deduction.". Beyond an acceptable sentence like that, I do not believe that this equivalence may be sustained. As a TLC engineer, fond of Relativity, I state that (Shannon's theorem) c is, in every instance, the speed of the quickest information medium available (to us, usually, the speed of light in a vacuum). If, say, tomorrow a gentleman from Mars lands here handling a walkie-talkie that uses "waves" 10 times quicker than our speed of light, from that moment c should become 10 times higher, and an atomic bomb would blast developing an energy 100 times higher.

Then, only people uncultivated in mathematics may maintain that it is impossible to make a trajectory, being in every moment with a speed less than c, with a global resulting time that suggests an higher speed; it is not the case of entering in mathematical details in here; suffice to remember that we are speaking about non-Euclidean geometry. Let me make a stupid example (often my students in engineering classes miss shamefully the answer): what is the shortest path between Roma and New York (roughly at the same latitude)? The first, obvious, and mistaken answer is a trajectory along the almost common parallel. Who knows why, on the contrary, commercial aircrafts, very keen to fuel consumption and flight times, prefer to head Northward, overflying France, nearing Greenland, then heading South-West along the shores of Canada: it's obviously the section of the shortest circle that connects the two cities on a spherical surface (a "geodetic", in relativistic terms). But, were I able to cheat (not really so!) I could break any record: were I able to open in front of me a straight tunnel between Fiumicino and JFK, an aircraft flying along it (under Atlantic Ocean, of course) wood exploit an even shorter flying time [27]. I repeat, it's a rather stupid example, but I usually get astonished when my students give me the first, awful, answer. Getting back to our 12-dimensions space (remember, a number also quoted in Ummites texts!), that is a Riemann metric, I'm always able, obviously from a theoretical point of view, to find shorter paths that allow me to reduce highly the chronotopic distance between two events, giving bystanders the impression that I must have moved quicker than c (which is not necessarily true).

Were then one to re-write Generalized Relativity in the terms I've stated (every entity is a complex one, quaternion analysis) one should find that the very same concept of "speed" is to be taken with great care: in our usual Euclidean space, with 3 dimensions plus time, speed is defined as the first derivative of space with respect of time (let's say 3 derivatives, because of the 3 space coordinates). From my point of view, I'd rather have 432 prime derivatives to take care of, and the three usual ones would be but 3 out of other 429 ones. I stop here, because, to get further on, I should enter details of calculus, and that is not the goal of a book like this one. I hope, any way, to have risen some perplexity in readers, and who knows if actually someone is going to take a pencil and a sheet of paper, to verify what I've stated.

What I may state to have apprehended (I repeat, within myself, may be under the maieutics by our friends) is that, in practise, we have to look critically to every phenomenon. I repeat, letting the W56's support me, it is necessary to be able to use one's thoughts, that are part of our entity, and refuse to accept at face value what our sciences and philosophy present us with. Another example: Cepheid stars. On these variable stars our astronomy has much invested in order to find distances from remote objects, assuming, without any evident reason, that some correlations were constant. Our so-called science is full of assumptions difficult to get justified, like the last one, starting from physics to astronomy, biology, and, *incredibile dictu*, to mathematics. Benoit Mandelbrot (he is not one of the W56's, but just a person with an open mind) started a critical exam of mathematics, anew, indulging in those border areas that classical mathematicians had avoided.

I repeat again, to the risk of getting annoying, that this has been the main teaching from the part of the W56's, on the subject of sciences. Both if my readers believe at what I've said (why should they, after all?) and if they believe that the three of us are as mad as a March hare, I invite them to take care of this suggestion. Be it from the W56's, or from your Author's ravings, nothing changes. It's maieutics.

Some Last Notes (Hans and Stefano)

A first thing that is worth to be underlined is that most of the languages of the Federation of the W56's do not use nouns, for grammatical reasons that would be too long to analyze here, therefore almost all of the names quoted here have been invented by earthlings. Therefore we have found improbable names like Gallarate, Romulus, or inflated names like Sigir, Sigis, Sirgis, and so on. It looks that we like Greek-looking names. An exception is Dimpietro, but I am not allowed to tell the origin of this name.

One thing the mean UFOlogist is not ready to accept is that "Flying saucers" and the like are not transportation media, or, at least, this is not their primary function: saucers, cigars, triangles, and all the other stuff, are not intended to transport people, but to act as mobile labs, may be even weapons, able to get where their assistance is needed.

It's easy to get convinced of that, if one looks to a generic witness report: were the object witnessed by a generic countryman in a generic field a flying thing, coming from a specific place and aiming to another specific place, we should expect other witnesses to report its coming and its going away: this happens very seldom. Most of times the object looks to have been created from nothing nearby, and having been annihilated again nearby (the W56's were maintaining that this is actually what takes place).

In one of my first lectures at the international symposia in San Marino [28], managed by the commendable Roberto Pinotti, I had presented (actually Roberto did, because at the very last minute I was unable to attend) a relation in which I was demonstrating statistically that UFO phenomena are almost always oriented to their witnesses. Only seldom it may happen that a generic layman, looking at the sky by chance, sees a scout that by chance was overflying him in that particular moment. Statistics suggest, on the contrary, that almost every episode is organized in function of whom, it had been stated, was to be involved in it. I suggest to UFOlogists to spend a few minutes of thinking on that.

What I believe the UFO phenomenon consists of, is what I call the "Conjuror Theory". Think to be in a theatre, assisting at the show of a clever conjuror: he will be able to exhibit unbelievable things. You are aware that there must be a well hidden intelligent trick behind all that, but our performer is really clever, so that no tricks may be perceived. Were you able to enter the stage, you might find that things are not like they were seen from the stall, but, unfortunately, you are not allowed to: you are allowed only to see just what your conjuror wants you to see.

In my opinion, the UFO phenomenon is just like that: somebody (something?) takes the care to put up an incredible stage, I repeat, one directly oriented to the one who has been selected to witness it. I do not hesitate to think that even the so-called aliens have been created to this purpose, therefore we find Ramu and friends who talk with Adamski on subjects our Polish mystic is ready to accept, Aura Rhanes that tells a Caterpillar worker (who obviously could understand nothing on the subject) of Clarion orbiting behind our moon, therefore not visible to us, and so on. All that, of course, filled with flying saucers, and various aliens, to give support to the tales. Of course, I put also myself, Bruno, and my other Earth friends, inside this stall.

I believe that this idea is particularly evident in the "abductions" stage; aside from the fact that, according to me, most of them derive from hysterical attitudes (in thousands of cases, never a single acceptable evidence has been found), the very few cases that look to have actually taken place present grotesque inconsistencies. Please think of the well-known case of Barney and Betty Hill, and think one moment of the "star map" that has been exhibited to Betty, and that has caused so many pseudo-contactees to start receiving messages from Zeta Reticuli. What could have been the use of such a map? If you think one moment, the only possible answer is to show it to Mrs. Hill.

It is sufficient to think a moment to realize that the so-called "Grays" cannot be stable living entities; just one consideration, among many more: if their eyes were actually spherical objects (as almost every kind of eye over this planet) they should intersect each other inside their head; the only alternative hypothesis is that they are ommatidia-based ocelli, typical of our insects, with a visual capability at very low levels (our insects typically use other sensors to get information on the environment).

The saga of "implants" is another example of how our conjuror is mocking his audience: the biological capabilities the W56's were able to reach, even after the death of physical bodies, we have seen, were astonishing, and they were able to interact at a molecular level. An item some mm long would have been unacceptably coarse to their eyes. Laughable, under similar considerations, the techniques used during the so-called medical analyses following abductions, of a level that our own medical science itself has left over since a long time.

It's therefore evident, from my point of view, that the abduction' saga (letting aside, I repeat, the majority of reported events, due to totally different causes) is but another trick put on stage by our conjuror.

And so on and so forth. Our conjuror (whatever may be concealing under this name) may use an evidently extremely advanced technology, in order to show us flying saucers, crop circles, animal mutilations, virgins of every kind, and whatever else. Of course there must be a reason for that. Any way, I believe that things, in this field, are far from what they look.

Such a criticism involves also the CTR's, the W56's, and the UTI's [29]. Both Hans and I believe to have got the proof that they were the same entities, who were changing habit according to circumstances. This implies that even the cosmic interlocutors in Amicizia were but one more trick by our conjuror. Honestly, I must state that Bruno rejected my idea, but I may understand his behaviour, because of the friendly *pathos* (this time in the conventional meaning), that was really involving. The W56's were friends to a topmost level. Their shows were wonderful, their culture in physics, technology, biology, and so on, was incredibly advanced with respect to our own. Any way, differently from Bruno, Hans was like me a bit suspicious.

Also their morality was more advanced than ours, to the point of not being as easy to get accepted. Try to suggest a judge that killing a man is not a crime, in itself, an action that could take someone in front of a court (see my talk at the Zanarini coffee-house). Yet, the very same idea may be found in the Bhagavad Gita (Mahabharata, VI), that is the base of yoga philosophy. Or try to tell the Pope, or anyone in the Vatican structure, that the morality is the one lived by friars (that Bruno loved so much, to the point of cutting any relation with a friar who, because of economic problems, had decided to become a priest).

The message our conjuror has been trying to convey, through the puppets named W56, CTR, UTI, Ummites, Elta V, and so on, had been many-sided: moral concepts, physic ideas, technology and biology, a diffused anxiety, together with other things that, I maintain, will have to remain concealed for long time, because our society is not ready at all to receive them. The centre for studies that Bruno had built, under direct suggestions by the W56's, remained at an initial state just because of faults from our part.

Any way, getting back to our conjuror, he does his (her/its) best to put up sceneries that are believable from a logical point of view, and so it has been during the long Amicizia saga.

I present here one of the many messages I've got, via "e-mail" (at those times e-mail as it is intended to-day was not existing, but there were similar systems):

PESCARA ARE BORING PEOPLE.
AGAIN LOOKING FOR THE SCOUT AT ORTONA UNDER SEA.
WONDERFUL SEA INDEED.
THEY NEED 300 MILLIONS LIRAS TO GET IT!!!!!!!
CRAZY LIKE DRUNKEN CRAZY CATS.
PAOLO STILL STUDYING TO BECOME PILOT.

HE IS ON SATELLITE NOW – GREETINGS FROM HIM.
BETTER NOT TO FLY WHE HE IS AROUND.
NO NEWS FROM MARS TROOPS. MAY BE LOST WHILE
DIGGING NEW CHANNELS TO GET WINE THERE.
GOOD LIVORNESE FISHES AT SERGIO'S.
HE TOLD ME HE KEEPS SPECIAL VODKAS FOR YOU.
YOU DIRTY CAPITALIST WHO STEALS VODKAS FROM POOR
POOR SOVJET WORKERS WAITING FOR COMMUNISM TO
OVERTAKE DIRTY WESTERN BUSINESSMEN.
ANY WAY, SPECIAL VODKA VERY GOOD, OTCIENI KHARASCIO.
THANK YOU FOR SUGGESTING SERGIO TO ALONE SPACE
WONDERER
FROM OUTER WORLDS.
SERGIO IS FINE, BUT WIFE IS EXCELLENT.
OUR FRIENDS AGAIN JUMPING UNDER ANCONA.
IF THEY GO ON, EXPECT ANOTHER MAJOR EARTHQUAKE BY
THERE.
BYE – SIGIS.

Aside a couple of Russian words (= "very good"), this message has been in English. Some comments for non-Italian readers: Ortona is a small town South of Pescara; Livorno (whence the adjective "livornese") is a town West of Pisa; Sergio's was a restaurant I was frequenting, while in Pisa; Ancona is the town where both Bruno and I have been born, some 150 km North of Pescara: at those times, Ancona had been struck by earthquakes during months.

Our friends were speaking Italian (often with a dialectal accent), English, German, French, Russian, but also classic Latin, Greek and Sanskrit (although, for practical reasons, Latin characters have always been used): IBM 2741 typewriters were not so easy to commutate from a character set to another! They often were changing their tongue inside a single sentence. For instance:

JDAIETE EVERYBODY NASHEGO DRUGA, KOTORIJ PRIJEKHAIET
TOMORROW NITE DA GIANCARLO.

(= "Wait you all for our friend, who is coming tomorrow night to Giancarlo's"); the sentence – who knows why, is a mix of English, Russian and Italian; evident problems arise in rendering Russian phonetics in Latin types; interesting the Americanism "nite" in the place of "night". Or even:

A VIJ SNAIETE CHTO APUTRA INSANIRE SOLENT.

That may probably be debatable, (= "You know that a person without offspring is prone to mental problems"), but any way it collects, in the order, Russian, Sanskrit and Latin, and a mistake: APUTRA is a singular name, while SOLENT is a plural verb; but may be it is right, if we consider APUTRA as a collective name.

Really amusing, at times, their Italian; they look to have learnt it in Abruzzo (the Italian region around Pescara), because they often were exhibiting a strong dialectal accent, and were often making errors in grammar: "Lui, e lui, andiedero…" (it means "they both went…", but with a gross mistake). A communication by Gallarate, via wireless, has got famous; this gentleman, Bruno has been speaking about in previous pages, came out saying that "…siamo protetti da razzi, proiettoli elettronici" (= "we are protected by rockets and electronic weapons"), where the word "proiettoli" should have been "proiettili".

What to say, at the end of this chapter? Obviously Bruno had never minded to convince anybody (when it was the case, facts were speaking by themselves), Hans and myself even less. I do even believe that, was there any proof to be exhibited, it should better remain concealed, so that readers may always think that this book is the result of the work by three paranoiacs (actually, with many other fuzzy-minded people with them).

Any way I hope (together with Bruno and Hans) that from one side this book may remain as the tale of an almost unlikely history, from the other part that it may help who is prone to, to open his/her mind towards concepts that we believe to be transcendental, and that on the contrary are immanent to ourselves, so that, he/she might be able to say, from Dante Alighieri:

> *E quindi uscimmo a <u>riveder</u> le stelle.*
> "and we came out to <u>see once more</u> the stars."
> (Inferno, XXXIV, 139)

I have underlined <u>*riveder*</u> because we are dealing with a rediscovery, finding something that we had forgotten, above all the being able to extract, from within oneself, what is quoted in the other famous verse [30]:

> *l'amor che move il sole e l'altre stelle.*
> "the love that moves the sun and the other stars."
> (Paradiso, XXIII, 145)

When the Adriatic sea went mad

At the end of 1978, in the whole central Adriatic sea, waters went crazy. It lasted a full couple of months. Huge columns of waters were rising from the sea all of a sudden, tens of meters high, unprecedented waves were running along the sea surface, strange lights were sighted at night, both civilian and military radars were receiving unexplained echoes. Let me present here just some of the hundreds of articles that in those days appeared on Italian newspapers:

The title reads: "Mysterious phenomena near St. Benedetto (a town some 70 kms North of Pescara): Columns of water rise from the sea all of a sudden".

LUCI VAGANTI RADAR IMPAZZITI COLONNE D'ACQUA

Adriatico misterioso

Un clamoroso caso di spionaggio o «finalmente» gli Ufo oppure strani fenomeni naturali? Il dilemma probabilmente resterà insoluto a meno che non vengano disposte esplorazioni più accurate del tratto di mare antistante Pescara. È proprio qui infatti che si sono verificati misteriosi avvistamenti di luci vaganti e di enormi colonne d'acqua da parte di pescatori e disturbi magnetici sui radar. Anche l'equipaggio di una motovedetta della Capitaneria di porto di Pescara, come forse sapete, ha avvistato qualcosa.

Il comandante della Capitaneria Gallerano ci ha dichiarato che effettivamente, nel corso di una perlustrazione in mare, un ufficiale (specialista in impianti radar), il comandate del natante CP 20 18 e i marinai dell'equipaggio, mentre erano in navigazione a 4,6 miglia al largo verso Nord rispetto all'abitato di Silvi, hanno effettuato un misterioso avvistamento che resta inspiegabile. Hanno visto un oggetto luminoso di colore rosso somigliante ad un razzo uscire dal mare e dirigersi a fortissima velocità verso l'alto, con un'inclinazione di 45 gradi circa, rispetto alla superficie marina. Nel frattempo il radar di bordo sembrava impazzire e si avvertivano forti disturbi magnetici. Come spiegare la cosa?

Qualcuno sostiene l'ipotesi del disco volante e non sarebbe questa la prima volta che nell'Adriatico si parla di misteriosi avvistamenti.

Più credibile (ma non tanto),appare l'altra ipotesi:nel nostro mare esisterebbero uno o più sommergibili di nazionalità indentificata, che svolgerebbero azione di spionaggio, magari con l'appoggio di un natante, mascherato da peschereccio, incaricato a sua volta di creare una specie di barriera magnetica al fine di non fare individuare il sommergibile in questione.

Ma quale scopo avrebbe ci chiediamo da parte di una unità incaricata di missioni spionistiche e che quindi dovrebbe occultarsi il più possibile, il lancio di un razzo?

Del resto il punto dove il misterioso avvistamento è stato fatto da parte degli uomini della Capitaneria di porto di Pescara, ha un'altezza di soli 23 metri d'acqua: impossibile quindi che vi si possa nascondere un sommergibile.

Sul posto inoltre la motovedetta ha fatto ampi giri di perlustrazione, ma non ha trovato proprio nulla. E allora? Forse si può trattare di mini-sommergibili.

Le domande a questo punto, come si vede, aumentano e l'atmosfera si fa quella di un film di fantascienza.

Per restare con i piedi un po' più sulla terra esaminiamo allora un'altra ipotesi quella fatta da alcuni geofisici e che, fino al momento in cui scriviamo, appare (sia pure con dei limiti), la più «ragionevole». Gli strani fenomeni verificatisi ultimamente in Adriatico sarebbero causati da fattori magnetici legati alle trivellazioni che sono state effettuate recentemente nella zona antistante le coste marchigiana e abruzzese per la ricerca di giacimenti di metano. Ma... (perché anche qui c'è il rovescio

Disco volante che sarebbe stato avvistato a Pescara nel dicembre del 1975.

della medaglia) com'è possibile che il «magnetismo» facia alzare in cielo meteore luminose o causi scintille?

Aumentano quindi le probabili spiegazioni, ma nessuna di esse appare ancora certa e scientificamente provabile. Il dilemma con ogni probabilità resterà a far parte dell'elenco dei tanti misteri che non da oggi circondano il mare.

Nicola De Gregorio

TEMPO 15-11-78

Niente UFO ma il radar capta un fascio luminoso

PESCARA, 14 — Le misteriose luci sono scomparse dall'Adriatico. Da qualche giorno non si segnalano più avvistamenti né da parte di pescatori né da parte delle motovedette della Capitaneria di porto che pure ogni notte escono a perlustrare le acque.

Eppure qualche fenomeno poco chiaro si ripete ancora: « I radar danno strane segnalazioni — ci dice Giuseppe Gasparroni —; mi è capitato di vedere sullo schermo comparire in direzione ovest un fascio luminoso analogo a quello che sta ad indicare la direzione della rotta. Non si tratta di un oggetto, perché altrimenti sullo schermo apparirebbe un punto. Invece compare una striscia luminosa che parte dal centro dello schermo, che coincide con il centro della nave, e che si estende fino al limite di "copertura" del radar. Non credo possa trattarsi di onde magnetiche che interferiscono, perché la bussola rimane ferma, mentre dovrebbe impazzire in presenza di un campo magnetico. Che siamo onde-radio? Anche quest'ipotesi però non mi sembra plausibile, perché le radio di bordo non accusano alcun disturbo ».

Here the title reads: "Mysterious Adriatic sea: wandering lights, radars gone mad, columns of water". The title of the smaller clip reads: "No UFO's, but a radar signals a beam of light".

TERRORE SUL MARE Nella ricostruzione del disegnatore Fabrizio Busticchi vediamo uno dei fenomeni che più hanno impressionato e impaurito i marinai della costa adriatica che va da San Benedetto del Tronto a Pescara. Immense colonne d'acqua, alte fino a quaranta-cinquanta metri, si sollevavano improvvisamente dal mare calmissimo.

"Terror on the sea"

Fishermen at first pretended to be escorted by the Coast Guard ships; but after the unexplainable sinking of a fishing boat (later it was found intact, sitting on the sea bottom, its two occupants died, but not drowned: the results of the autopsies conducted on their bodies were never released), fishing activity stopped almost completely.

In the meantime, over land and sea, a veritable flap took place, with hundreds of UFO's and occupants seen by every kind of people, from Policemen to farmers. Just to show how pervasive was the UFO activity in those days, let me present here a cartoon that appeared on a major Italian newspaper (Il Tempo), among many articles on the subject; its caption reads: "It is the 23:48 UFO".

A Police officer in Chieti stated, during a radio broadcast, that he has sighted at night… "An UFO as large as a three-floors building!"; it was so luminous that he and his men had to wear sun spectacles to be able to look at it.

A farmer noticed a "spaceship" resting on the ground, was able to approach, and saw men and women inside it, who were apparently mocking at him! An article told of another farmer, who was said to have pursued an alien, menacing him with his pitchfork, but the man (a very rude one) later on denied this last detail, stating that, on the contrary, he fled away, on being confronted with the mysterious apparition.

Of course electrical black-outs were common, one of them has been really unusual. On the 26th of December, workers inside a small power plant near Pietra Camela, on the cliffs of the Gran Sasso mountain, sighted a luminous sphere just outside the plant. The generators went immediately dead, but an auxiliary unit, which has been shut down since months before, started by itself!

No rational explanation has ever been found. Astronomers at first referred to Venus and Jupiter (that actually in those days were very shiny), but of course these planets could have caused just a few of the sightings. Someone spoke about a secret military activity, may be submarines from the Soviet Union, but no evidence emerged (a rather risky adventure, if true: the eastern coasts of Italy were an important NATO border). The situation was so heavy that a commission of study was formed by the countries on both sides of the Adriatic sea, but its conclusions were never made public.

Your Author has been interviewed on the national broadcast, but he decided it better to "play cool", so agreed with the astronomical explanation. The following graph is the result of a FORTRAN program I've written to evaluate the intensity distribution of the phenomenon:

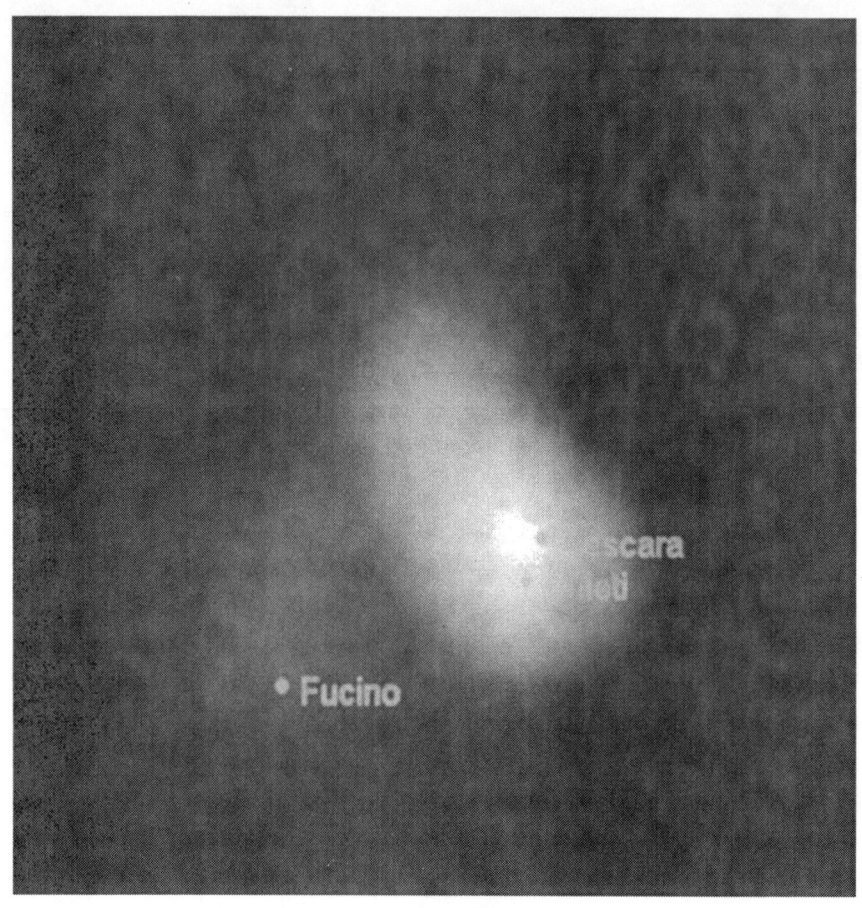

By the way, in these very days (August, 2008), fishermen start reporting once more that they see strange lights beyond the surface of the sea, at night; in the meantime, scouts are being spotted here and there around Pescara...

Conclusions

What has been written is not meant to be an exhaustive description of Amicizia. Of course I can't say to have learnt everything on the phenomenon, far from that! Bruno himself, when we decided to write down something on this story, was knowing, outside Italy, only something in Switzerland and Austria, but he couldn't imagine something having taken place in other countries, some of which has been later told him by Hans. Who knows if there is someone who actually owns the knowledge of what Amicizia has meant, *in toto*? Who knows how many persons, all over Central Europe, are acquainted with it? Of course the best suited one had been Dimpietro and the other puppets designed by our conjuror.

But, if to our sorcerer the operation requires just a bit of his immense technology, those who are the objectives of his activity get strongly affected in their everyday life, often get exalted by this new adventure, and jump enthusiastically into it, without being aware of the risks they're running. The game is by far extremely rougher than any of us is able not only to afford, but not even to tolerate.

Among the many reasons because I was not keen to write down such things, there is actually the risk that some simpleton, moved by our tales, would try to emulate ourselves. To anyone nourishing such ideas, I underline that Hans himself had slammed the door in front of the presumed aliens, just because there was no sense in going on. I repeat, too many persons have died within this history, even people foreign to it, too many persons went mad, too many have ruined their life. And for what all that? Just to have been in touch with a reality that, although immanent, is any way too transcendent to our daily way of life? Has it been worth of it? I don't think so.

I know a few people who have voluntarily stopped their involvement; the large majority is rather satisfied of what has been going on (and in some case is still going on). Probably they are short-sighted, because they do not care to investigate the meaning of all that: why aliens should have decided to meet Howard Menger? Just in order to have him enjoying the reactions of their females when confronted with bras and spiked heels, that he had been clever enough to provide them ("From Outer Space to you", Saucerian, 1959, pg. 71,72)?

Moreover, contact activity is always originated by the presumed aliens. There is no other way to get in touch with them, if not indulging in rather dangerous activities (for instance the common "sky watches", that are dangerous because no one knows who, or what, is going to answer, if he/she/it will want to). And usually the contact starts with a period of hidden overseeing, that lasts for a very long time, before actually a contact takes place.

It is unavoidable to ask a lot of questions. For instance, why, within Amicizia, a psychiatrist, two cardiologists, a diplomat, an archaeologist, some twenty engineers, several accountants, an expert in military logistic, bank employs, two members of the FAO, five university professors, two managing directors of important Italian companies, a jewels wholesaler, a judo champion, a Court of Assizes judge, the Executive Vice President of one of the largest multi-national companies all over the world, one (or may be two) future Nobel laureates (in disciplines far away from one another), four generals (three of them Italian, the other an Austrian gentleman), plus a hundred of people (some politicians among them) have been involved? Why, in the meantime, UFO buffs have been taken outside (probably Perego has been the lonely exception)? This last sentence, actually, is not totally true. In Italy some relevant person in the UFO environment have been contacted, most of times without even realizing that, but without any peculiar goal, just for the fun of it.

Why no church prelate in the previous long list, no friar, although Bruno was very fond of them? Probably it's a mistake of mine, deriving for my un-perfect knowledge of things within Amicizia, because it's too strange to be believed, and, as usual, some signs would make us think the other way.

During my talk with him in the Zanarini coffee-house, my friend had presented me with some rather obvious explications (mental strength and the like), but it is evident that these are but formal reasons. On the long run I got convinced that it has been just a transferring of technological and moral information, with the W56/CTR/UTI scenario put up to give a sense to the thing. Different scenarios have been put up elsewhere (for instance in the States, with the very same conjuror acting), although none of them, as far as I know, has been as showy, as long, as widespread as ours. Among other things, the circuit for TLC amplifiers I've spoken about before is but one result of such a technology transfer.

Summing up, and going to end, I believe that what I have written must be taken *cum grano salis*, and I'll not be the one to resent, if some reader will think that we are totally mad. After all, I'd even suggest such a proposition, for the reasons I've already underlined. And, beyond Amicizia, I'd like to take advantage of the occasion given me by these pages to warmly dissuade anybody from trying to approach *de visu* the UFO phenomenon: it's of no use, and it is usually dangerous, both because of eventual CTR reprisals (inside the Amicizia saga), and because any way we are not ready for it: it's too easy to find oneself out of place in one's environment, after having received information from beings so much more advanced (be they real or not, it's totally meaningless).

So, let people go on investigating, if they want to, small lights in the sky, or a picture of a scout taken elsewhere: it's an harmless activity (and an useless one, in my opinion).

But do not try to get a direct approach with this fictitious reality.

Gaspare De Lama

A few weeks after this book (Italian version) had entered bookshops, I received a very polite letter by Mr. Gaspare De Lama (among many more else, he had shot the picture no. 52, presented in the photo section). The painter, now 85, is the person quoted here and there in Bruno's memories, just like "Gaspare". He was congratulating me for the book, protesting just for a sentence of mine; a few pages ago, at a certain point I wrote: "Has it been worth of it? I don't think so."; Gaspare wrote me: "I'd have written: I do think so!".

So we got in touch (I had never met him before, although, through Giancarlo and Bruno, I knew a lot of things on him), and became friends; in a lecture I had been asked to give about the book, Gaspare attended, and I asked him to say some words about his experiences. Later on, I asked him to write down something, and he sent me a few hand-written pages that, with his permission, have been published in the major Italian UFO magazine, "UFO Notiziario", and that I like to translate here.

Gaspare's Report

The beginnings

I was fond of UFO's since 1948, I had read one of Perego's books. I wrote him to express my opinion, and the consul phoned me back; later on he came to Milano to meet me (I was living in 19, Giulio Cesare square). We became friends (I still have a book of his, that he signed for me). We met many times, and he offered me to become the secretary of CISAER (Author: A private group of UFO scholars, founded by Perego). I declined, to remain dedicated to my painting; moreover I should have gone to live in Rome. He never told me about the W56's.

One day I saw, in a magazine, flying saucers pictures, within an article signed by Ghibaudi (Author: This journalist is the "other Bruno" quoted at length before). I wrote him a letter, he phoned me back, and within a few days he came to meet me. During the conversation, he told me how things actually were. Later on he came again, this time with Bruno (Author: Sammaciccia), who, this way, got in touch with me and my family.

Mirella, my wife, was very interested, the same with my mother (then about 70). One day my mother told me that Sammaciccia had come to my house when I was out; she and Bruno remained talking for a while; he had counselled her to push me to give up, because the W56 story could become dangerous for the quietness of my family. My mother answered that I was of age, that she didn't want to interfere in my decisions, and that she would have agreed anything I was going to do.

The present

Few days later, a night Bruno phoned me (he was already living in Milano, in Capecelatro street, at those times in the outskirts of the town). He came, and we got out on the street, in order to see a disk that was to pass.

Actually it came, at a very low height (we thought some 400 metres); it was larger than the full moon, of a pale orange, shiny, but with definite contours. It was 11 p.m., and nobody was in the street.

Later on, Bruno told me that the W56's had decided to "present" me. I was to chose a place in the country around Milano (without electric power pylons on it), then to buy an 8 mm film. Next day Bruno asked me to take him to the place I had chosen (in the Trenno area), and asked me to insert myself the film into his camera, and started to shoot at an empty sky, while I was near him. He took also some still pictures. When he was over, he asked me to take away the films, and have them developed; next day I went to the Ferrania, in Matteotti street (Author: A well known Italian films company). When I got the developed film, I took it home, and my family and I appreciated it. It lasts a few minutes, and I still have it, although it has been a bit damaged with the time. Some of the still pictures Bruno shot in that occasion are actually those presented in the book "Contattismi di massa", from no. 16 up to no. 25, plus some one else, that I still have.

Author: Let me comment this strange story: when I have read it, at first I didn't believe it, for a very good reason, that I am going to explain. At the end of the 60's, I shot some 20 pictures of a low-flying scout, that actually landed, near Montesilvano (North of Pescara); it was very early in the morning, a misty day; I was in a rather rectangular clearance, surrounded by trees all around. The scout came in from the West, flew a couple of times over the clearance, at tree-top level, tried to land on the North-East corner of the rectangle, then, at the very last instant, its pilot realized that the soil was very rough in that point, rose again, and finally landed on the South-East corner. Later on I gave Bruno both the stamps and their negatives, so that he could distribute them among his friends. When we decided to write down this book, I asked Bruno to give me the pictures he had, in order to insert them into it, but he answered that he had been left with very few of them! Giving them around with liberality, he was now left with very few pictures (at Amicizia times, I had seen literally hundreds of pictures, at Bruno's). So he started looking for pictures among his friends, and a Swiss lady agreed to send us, with a parcel service, the pictures she was keeping in a bank vault (!), pretending that we sent them back as soon as possible in the same way; among these pictures, the ones I present in the photo section, nos. 40÷49, and I was sure they were my pictures; while looking at them, Bruno said something like "treno", that I understood to be the Italian word for "train". I could not see what a train was about with these pictures, but as we were in a hurry, I didn't care to ask him. When Gaspare made his previous statement, I didn't believe him, because I was sure that I had shot those pictures. Then Gaspare lent me the movie, and I had to realize that pictures and movie were looking to be one and the same thing.

Not yet content, I went on with orthographic analysis on all the pictures, via computer, of course, I have been teaching for years Computer Graphics! So I discovered that they show two different objects; the following drawing presents the shapes of the two scouts, reduced to a same diameter:

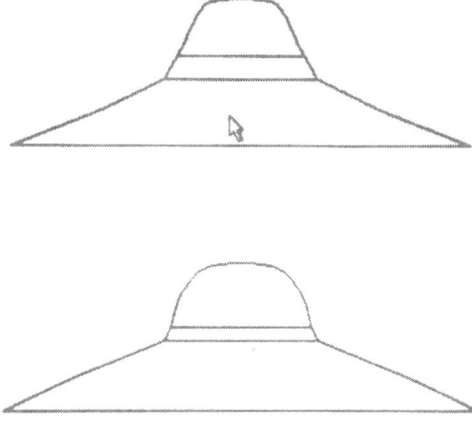

Evidently they are slightly different from one another, so, probably, only the pictures no. 42, 44, and 49 are of mine (the lower shape), while the others are connected to the Trenno (not "treno", as I had understood!) episode. Probably, while the pictures were being passed from hand to hand, someone made a muddle of the two sequences, loosing many pictures in the process, and what the Swiss lady keeps in a vault is actually the mix of two different sets. The surroundings are very similar, therefore it is not so strange that a mixing occurred, but it is symptomatic of the small care taken in preserving such memories. Now that I've explained, I hope, the *qui pro quo*, let's get back to Gaspare:

The phenomena

Both in Milano and Pescara, in Bruno's house in Genova street, I (and often my wife with me) was with Bruno, and really often I have witnesses the following phenomena:

In Montesilvano I have seen (always upon appointment) a disk in three different times, with unambiguous movements; in one instance it was moving like that:

In all evidence, it was making a show!

I've witnessed very many times objects being created from nothing, as if the final stage of a transmitting process; at times they were appearing near the ceiling, then falling down, other times they were appearing on the ground, then rising quickly, knocking on our chairs. Most of times they were movies, or tape recordings with the voices of some W56's in there (Sigir most times), who were giving us some directive, or just cheering us. When I was taking these objects, they were always tepid.

At times, blue flashes (2÷3 metres long) were appearing and going zigzagging on the floor, at times passing near to us, but never hitting us. We had been told that they were to put, or to verify, control and defence systems inside the house.

I've witnessed the phenomenon of street lights switching themselves off and then on again when we were passing under them (both walking, or in a car).

One day Bruno asked my mother, my wife and me to look through the eye-piece of a camera, aimed to the sky. "Do you see anything?" he was asking us, but there was nothing to see; then he would say "Look again now", and the camera was now showing a flying saucer; then it would disappear, and another one, may be a larger one, came in its place… It was an amusing show.

I've seen Bruno make telephone calls, with numbers suggested by the W56's, numbers 20 figures long (Author: At those times, with electro-mechanical switches, this was a plain impossibility, but I too have tested it, in my case numbers were 30÷35 figures long!), getting connected without passing through telephone switches, and free of charge.

Once I was asked to act as an "aerial" (just to give it a name), I do not remember for what purpose, over the base in Pineto, at night. I was alone, inside a wood, and from time to time flames 30 cm high were lighting up and switching off all around me, some 2 metres away (I was a bit perplexed…), but Sigir, via wireless, told me not to fear anything from the CTR's. Speaking about radio sets, the one I had had been arranged by the W56's so that, while receiving their messages, changing the tune was not affecting the reception.

Another time, I, together with other persons of our group, were asked to wear pendants down on our necks: they had been made in wood, with small copper nails into them, following their instructions. Bruno had a telepathic conversation with some of the W56's and some guys from another race, friends to them; all that lasted half an hour. At the end, we were asked to take our pendants away, and, to our astonishment, we found that the side that had been in touch with our chests was carbonized, although our skins had suffered no damage!

Once Sigir cheered all of us via wireless (we were sitting in circle, my wife was with me), calling each one by name, in the very same order we were sitting, as if he was able to see us; I believe that he was actually looking at us.

One of the best memory I have: I've been astonished (at that point nothing was able any longer to amaze us) because of the poetry exhibited. I was standing, leaning against the trunk of a pinaster, on the hill near to Pineto, and, through my small radio set, I was in touch with Sigir. It was a sunny afternoon, and I was there to fulfil some of their purposes. Sigir was speaking to me, and at a certain moment he told: "I know, dear Gaspare, that you'd like to be down with us, and embrace us, but you can't imagine how much I'd like to be up there with you, outside this base, in the open under the sun, under that pinaster, where now a small bird is arriving"; actually I saw a small sparrow flying toward the tree, then stopping on a branch a little over my head "and now it will start to chirrup" and it started "and now it will happily fly away" – and it flew away toward the sea. I'll never forget that episode.

The End

In Milano I got acquainted also with Giulio, who had been for some time in San Salvador. We became friends, I was feeling well with him. Unfortunately, he went away, after a furious quarrel with Bruno (I know the reasons, but it would be too long to discuss them here). That evening, Raffaella (Bruno's wife) phoned me in agitation, begging me to rush to their home. When I got there, Giulio had already fled away, being menaced by Bruno, Raffaella told me that Bruno was to break a bottle over his head (questioning our Friends was making Bruno go wild…). So I went to Giulio's hotel, and found him upset and weeping. Their friendship was over. Then he went away, and I haven't seen him since them.

One day, we were alone, Bruno and I, Sigir sent us a wireless message; he was complaining because an "Higher Entity" (UTI, may be?) had harshly censured what the W56's had done. Very severe were the words of this censure. Bruno found them unjust, and that has been the only time I've seen him weeping (but for anger); I try to calm him down, I couldn't see him in that state.

Battles between the CTR's and the W56's were now a daily affair; moreover Bruno was exasperated because of the way things were going on inside our group. Unfortunately I was noticing that Bruno himself had started to back bite someone of the group, and to relate me unfair words said against me by others from within the group. I was not understanding this totally new behaviour, after our Friends had asked us to try to stay together, in harmony, elsewhere Uredda was to corrupt itself. I was deluded, disconcerted. Then, once Bruno behaved in a really inhuman way against the architect Walter and his wife Verena (as with Emilio, it had been myself to introduce them into the group); Bruno was wrong, from my opinion, but my intercession was to no avail. He dared to rage on them for more than a month (although Walter had done so much for our cause, and Verena was the most beautiful woman – I mean from an interior point of view – I have ever met). Moreover, I repeat, their fault was a minimal one, and unintentional. I was astonished (the W56's had nothing to say on that?).

Other things took place, not so serious, but many of them, and I was affected: I was no longer liking such a Bruno. He was really trusting me, and often was saying "I'm sure that you'll never betray me", but while he was saying so, something had changed within myself, I was deluded by the new Bruno, although I was trying my best to understand his reasons. I had always a kind of love to him, but I was willing to be honest to him, and, above all, toward me. Therefore, I decided to get out.

Via the judge B. (as Stefano has quoted him) Bruno sent me the tapes recorded by our Friends, who were soliciting me to get back, with moving words. I didn't.. Bruno wrote me many letters, very lovely ones, which I always answered; I still keep his letters, and the ones by Verena, who once described me how our friends had healed, in an almost miraculous way, her daughter Maja, after an abortion she had suffered.

Years later Bruno came to Milano (he had no longer his house there), phoned me, I went to his hotel, and we had lunch together. I tried to explain him that I still was loving him, that I was grateful for the experiences I could never had had elsewhere, if not for him (I was true). He was looking at me with a reproach in his eyes, and I understood that I was to see him no longer. But I still have Bruno in my hearth.

Final remarks

I could tell so many other events, and phenomena, that I remember quite well… For instance, the time Bruno told me that the name W56, beyond referring to the starting date of the story, was also meant as an homage to George Washington, who had been the only President of an

Earth nation to have got some contacts (I don't know in which way) with the W56's. But I repeat, whatever I could add, would not change too much the history of the W56's.

Post scriptum

My involvement with the W56's lasted from 1960 to 1965.

Amicizia still going on

Although, according to Bruno, Amicizia got to an abrupt stop some thirty years ago, actually I believe that the story never ended; it did for Bruno and for his group, it did in Germany too, but nevertheless something continued to go on, here and there; then, in the last years, our friends started again to walk our streets, to get in touch with new people; it is not strange that they decided to look for fresh mates, because the previous ones had now grown too old, if not even passed away. Here, in this last page written just minutes before sending this text to the printer, I summarize some of the most remarkable events that have taken place in recent times. Obviously, this list is far from exhaustive, both because I don't have the time and because I do not want to extend too much the number of pages of this book.

A few years ago, in summer, a friend of mine, 1.94 metres tall, was strolling near the University of Pescara, with some of his friends, around midnight; they entered a square, overcrowded, so they decided to walk around its perimeter; in doing so, my friend noticed three persons, two boys and a girl, resting in front of a wall, without any window or door in it; when he passed near to these three persons, my friend, used since ever to look downward at people, was upset, really frightened to have to look upward! The girl was roughly as tall as him, but the boys were some 2.5 metres tall. Their height was unnoticed to the people around, probably because of the lack of reference points (windows and doors) in the wall.

This spring (2008), just one week before a convention was held in Montesilvano, where Gaspare De Lama, Paolo Di Girolamo, and myself, were, for the first time, to relate on Amicizia, another friend of mine was walking near the Tiburtina station, in Rome (it's the second main railway station in Rome) with his wife; he saw a men walking towards them, 2.5 metres tall; he walked after them, apparently unnoticed by the other people present in there. My friend, who knew almost nothing about UFOs, then attended our convention, learnt about Amicizia and the tall people involved in it, and told me about his unusual encounter.

Recently, other very tall people have been seen in Milan. It's strange that, in these cases, people around seemed to be not worried by the height of these unusual persons.

Then, there is an activity taking place since some years in Southern Chile, this time under the unusual name "Friedenship", with aliens, presenting themselves with the names of the major angels, interacting quietly with the community of the small island they live in, making contacts via wireless with other people in Chile, and even making broadcasts, on a roughly regular basis, over short waves.

Here in Italy, a gentleman from Vasto (some 70 km South of Pescara), and a professor from the University of Florence, maintain they are in touch with aliens, who look very much like our W56's. Of course, I do not know whether other people are involved. Strange episodes are taking place in Veio (North of Rome; refer to the Ummo chapter). Then, I repeat, the Adriatic sea seems to be once again perplexing fishermen, with lights appearing below water level (in a few days some friends and myself will sail, on a fishermen's boat, to see what is taking place).

So, as we say in Italian, nothing new under the sun!

Appendix

Before, I've quoted a theorem, according to which all triangles are equilateral. Probably I have provoked some perplexity, so it's the time to resent it.

Let's take a generic triangle, ABC; let's draw the bisector of the angle in C, and the axis of the segment AB; be E the middle point of AB, and D the intersection between the bisector and the axis. Let's connect D with A and with B; from D let's draw the perpendiculars to AC and BC:

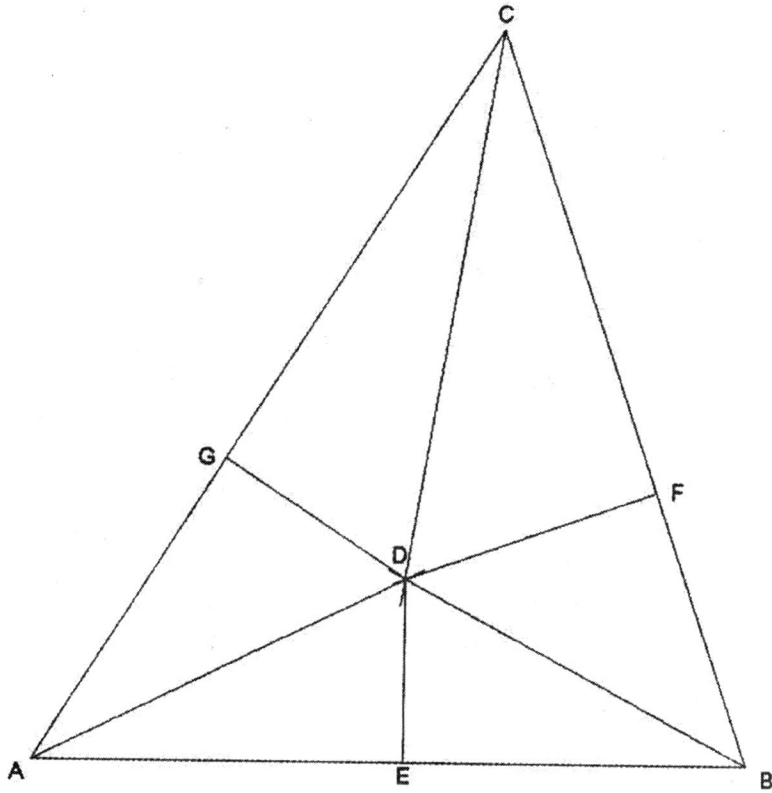

The triangles CDG and CDF are equal (both are right-angled, have the same hypotenuse, and equal angles in C); therefore DG equals DF and CG equals CF.

The triangles ADE and BDE are equal (right-angled, two cateti orderly equal); therefore the hypotenuses AD and BD are equal.

The triangles ADG and BDF are equal (right-angled, hypotenuse and a catetus orderly equal); therefore AG is equal to BF.

The sides AC and BC are therefore equal to each other, because they are sums of equal segments, and so the triangle is isosceles on the base AB; iterating the process starting from, say, A, the triangle will result isosceles on the base BC, therefore it is equilateral.

I repeat, this demonstration is formally correct (from Euclid's point of view). The problem consists in that Euclid have never given a formal definition of "inside" and "outside"- To-day, what I have presented is the proof ab absurdo that the point D cannot be intern to the triangle ABC (were D inside ABC, then ABC is equiteral!).

Photo Section

Some comments on the following images, presented here in order to underline what has been said up to now. Speaking of pictures that show something strange in the air, of course a picture doesn't prove anything: it's too easy to realize plausible fakes, if one is able to – see picture no. 64 in this section. It must be noted, any way, that most of these photos date back to tens of years ago, when the technique, and above all the computer peripherals available were decidedly rudimentary ones. Moreover, many photos are Polaroid ones, really difficult to fake (although it is not impossible, good skill is required, together with a lot of money).

01: George Adamski;

02: George Hunt Williamson;

STATE OF ARIZONA)
) ss.   AFFIDAVIT
County of Navajo)

We the undersigned, being first duly sworn do solemnly swear; that the Documentary Report of Interstellar Communication by Radiotelegraphy entitled: "THE SAUCERS SPEAK", is accurate and true. We have been witnesses to and participants in these happenings as listed in the above report. We also state that we are trained observers by the very nature of our various occupations. We agree that a fact is not a fact until it is proven; and that much time, effort and research has gone into the above report to prove beyond any doubt its statements to be absolute fact. Our work was carried on in the acceptable and standard form of scientific research employed by radio operators, anthropologists, and others. We also state that many tests were performed under exacting conditions.

We are thoroughly convinced that the "Flying Saucer" phenomena is interplanetary in origin; that the mission of these ships to our Earth is a friendly one; that the "saucer" intelligences have developed ESP (Extra-Sensory Perception) to a high degree; and that these intelligences are of the human race inhabiting other heavenly bodies and are now attempting contact with any inhabitants of the planet Earth who are receptive to the Universal Truth.

We also swear that we are not members of any organization (religious, scientific, etc.) that would in any way profit or gain from our research. We are not propagating a dogma or creed and none of us involved would gain by perpetrating a hoax.

This report is being given to the people of the world because the facts contained therein were not given to us for our own elucidation, but are for all those seeking and desiring Universal Truth.

That the undersigned Mr. S, on oath deposes; that he received the messages in International Morse Code.

Mr. S, Licensed Commercial and Amateur
Radio Operator.

Licensed Radio Operator

Mrs. S
Winslow, Arizona
Housewife

George H. Williamson, Sc.D.
Prescott, Arizona
Anthropologist

Betty Bowen
Winslow, Arizona
Student

Betty J. Williamson, B.S. A.B.
Prescott, Arizona
Chemist and Anthropologist

Ronald Tucker
Winslow, Arizona
Student

Alfred C. Bailey
Winslow, Arizona
Conductor, Santa Fe Railroad

Betty M. Bailey
Winslow, Arizona
Housewife

Subscribed and sworn to before me this 7th day of March, 1953.

Notary Public
Winslow, Arizona

My Commission Expires 10/26/56

03: The affidavit signed in order to confirm what is told in "The Saucers Speak!"; it is the original version, taken from the manuscript of the book;

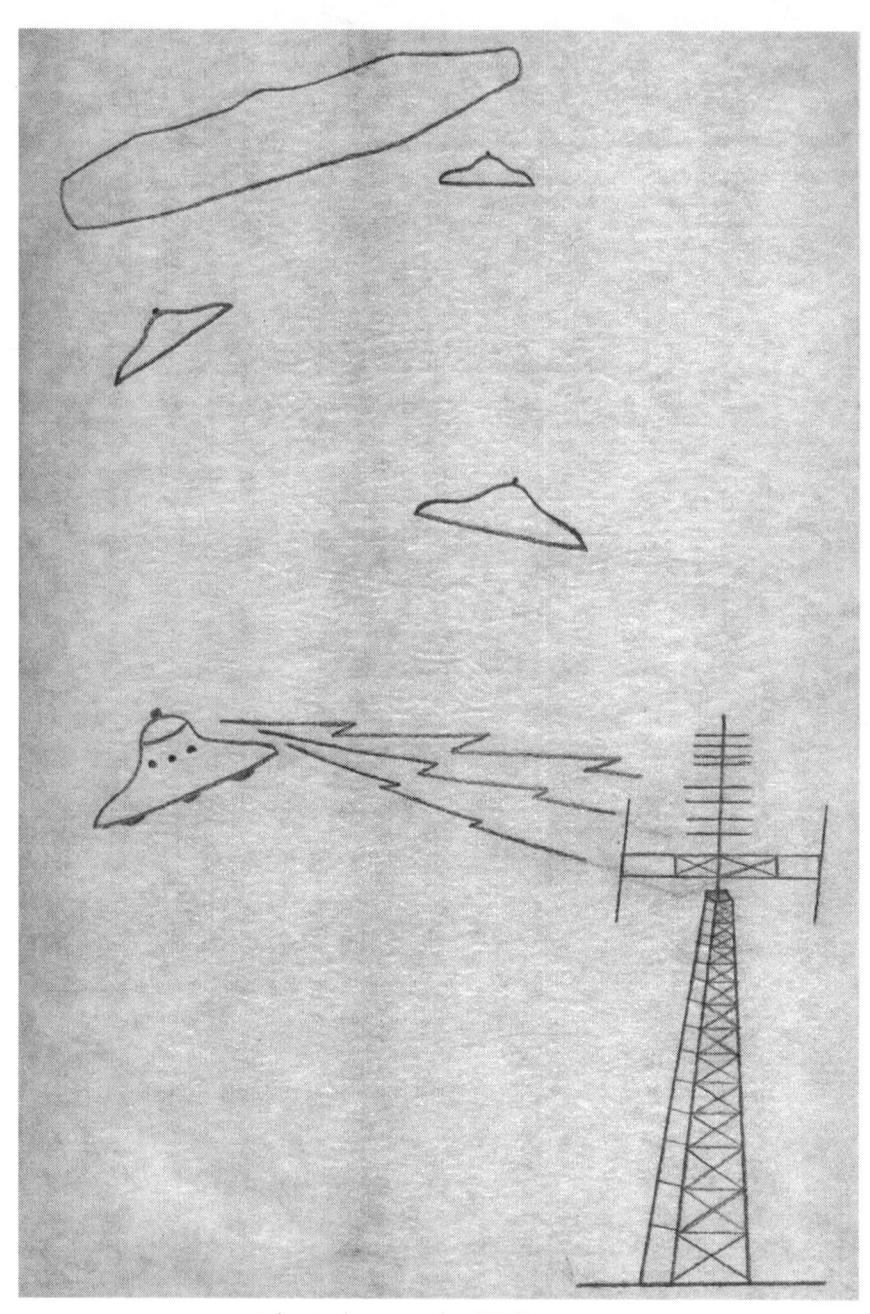

04: A drawing by Williamson,
that they decided not to include inside the book;

05: Daniel Fry;

06: Desmond Leslie, together with Wendelle Stevens;

07: Howard Menger;

08: Truman Bethurum;

09: Aura Rhanes, who was in charge of the flying saucer from Clarion;

10: Who says that the aliens living among us do not like to get pictures taken? Here we have the prince Neosom, together with the princess Neoganna, from the planet Tythan. After this picture has been taken, the prince's father migrated to an upper level of consciousness, therefore now the former prince has become the king of Thytan; the couple frequented the annual convention at Giant Rock;

11: And, from one of these conventions, a dog from Venus; its names was Queenie (this picture and the former one are taken from "Flying Saucers and the Scriptures").

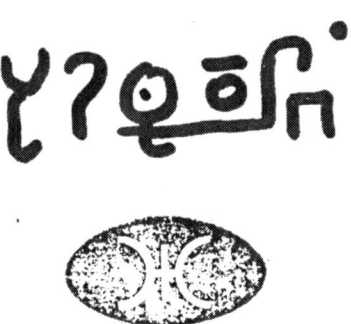

12: An exemple of Ummite letters.

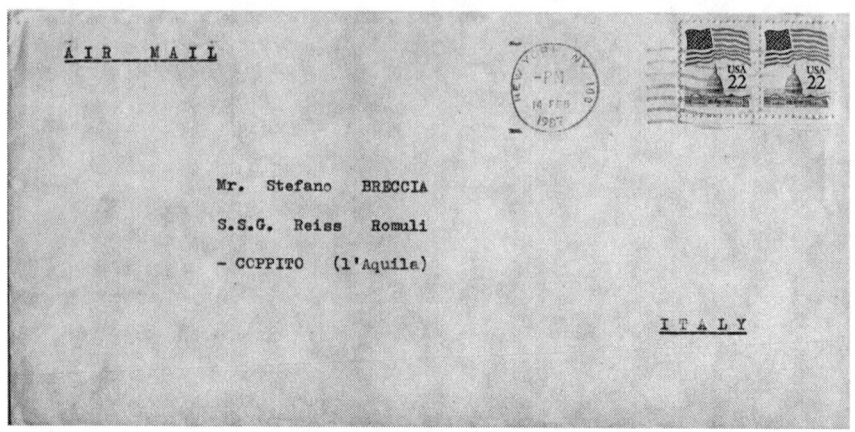

13: The envelope of a letter sent to the Author from a post office in Manhattan, NY; it must be noted that, the same day, from the same post office, 2,000 letters were sent, all equal to each other, addressed to people all over the world. It must be noted that the company I was working for was "Scuola Superiore Guglielmo Reiss Romoli", and that the last word is mis-spelled.

14: An Ummite symbol, found by my friend Carlo Bolla, carved inside a grotto in the necropolis of Veio.

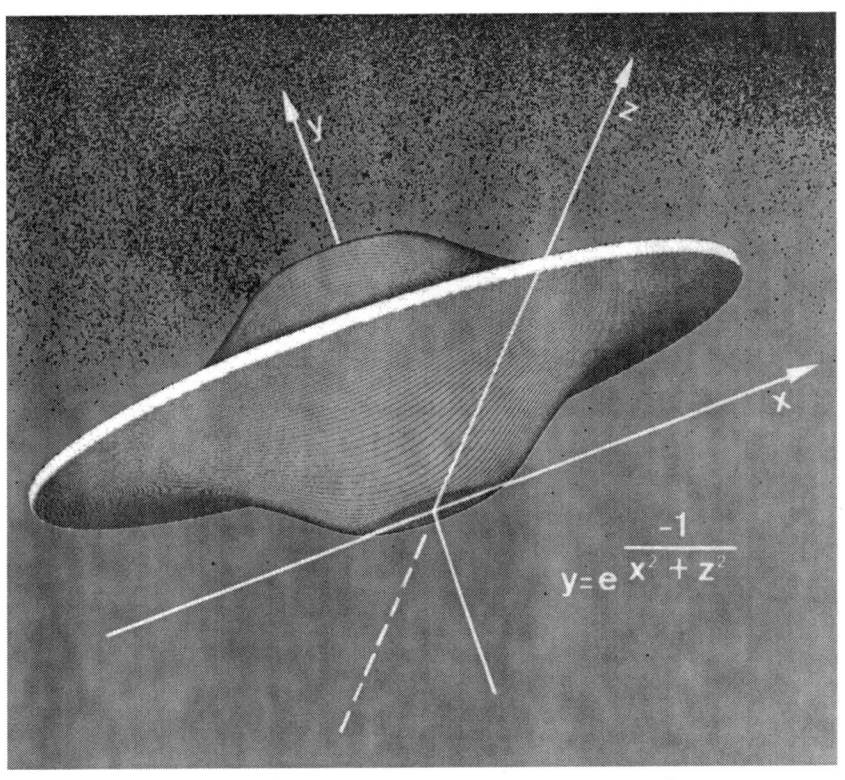

15: A computer-generated synthetic image,
showing the optimal lenticular profile, according to Ummite physics
(this image is from the cover plate of "El Pluricosmos").

16÷17: Sequence of the landing at San José de Valderas; pictures no. 1 and 2 among the ones shot by the photographer named X.

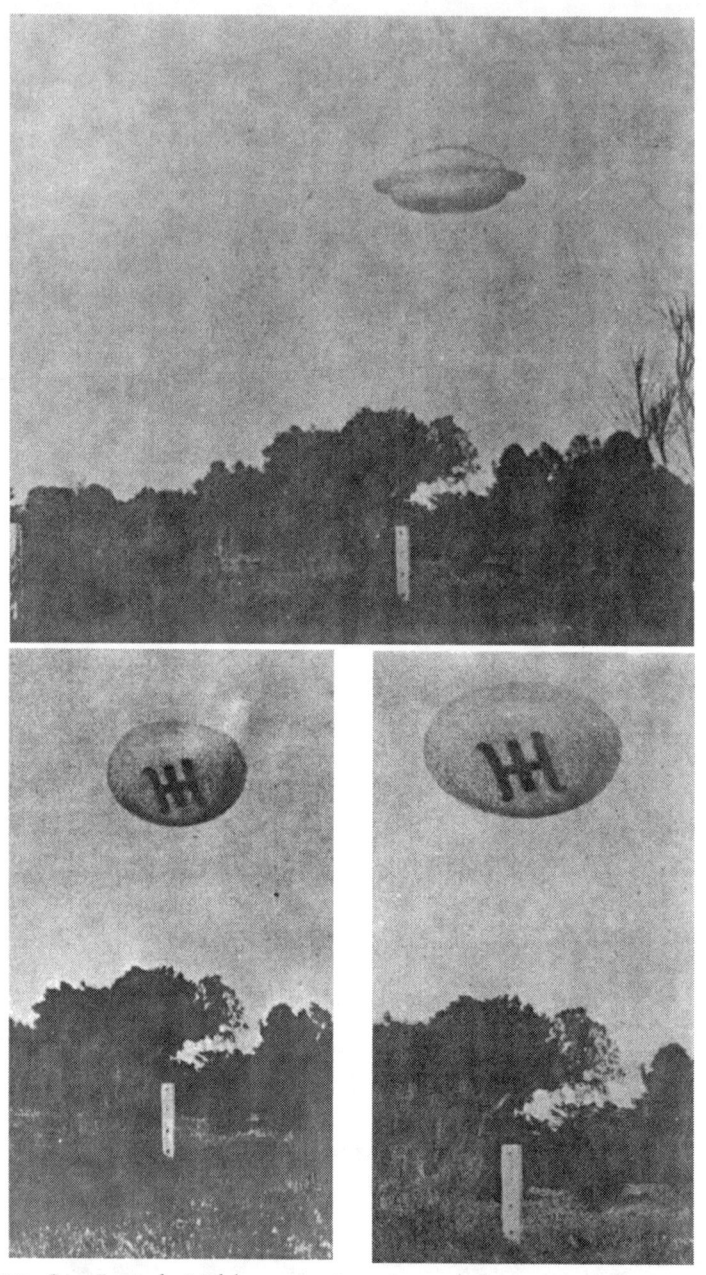

18÷22: San José de Valderas again; pictures from no. 1 to 5 among the ones shot by the photographer named Y.

23: One of the marks left by the landing tripod; from its depth a weight around 15 tons has been estimated.

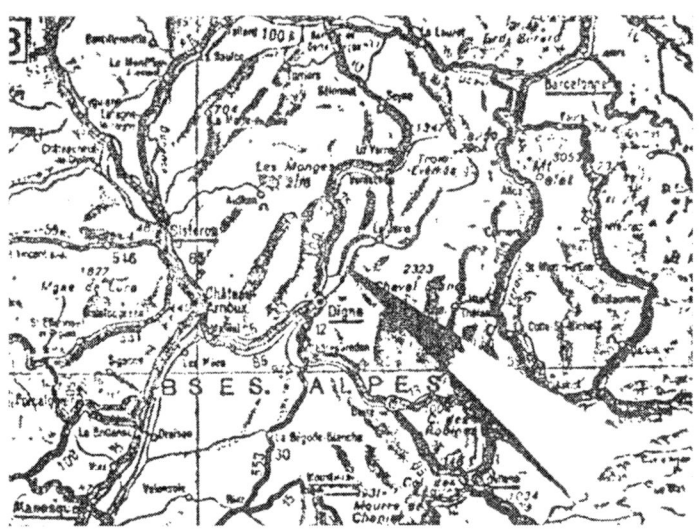

24: The location of their first landing, in France.

25: The gentleman in the centre is Bruno Sammaciccia, to his right, in the foreground, Giancarlo; the picture has been shot in occasion of the inauguration of the big villa, in Montesilvano. The other people are their wives, a General of the Carabinieri, one of the Army, and the former director of the TIMO (the telephone company that was in charge of Central-Eastern Italy).

26: Bruno Sammaciccia together with the Cardinal Agnelo Rossi.

27: Bruno Sammaciccia with the General of Carabinieri, Gaetano Tamborrino Orsini.

28: An autograph of the former Archbishop of Cracow, as a dedication to the book *Le Miracle Eucharistique de Lanciano*.

29: "Annunciation" by Carlo Crivelli (1430÷1495); in this strange painting, the Holy Virgin is kneeling inside a closed room in the lower right, and is reached by a beam coming from a circular object high in the sky; this UFO *ante litteram* has a well defined structure.

30: "Nativity" by the Ghirlandaio (1449÷1494); while Mary is adoring her little Baby, a bearded man (probably St. Joseph) is looking at a globular dark object high in the sky.

31: "Bacchus and Ariadne" by Tiziano Vecellio (born in 1488); on the top left section, half concealed behind some clouds, several bright objects are arranged in a circular pattern, and apparently they go unnoticed by the many persons present in the scene.

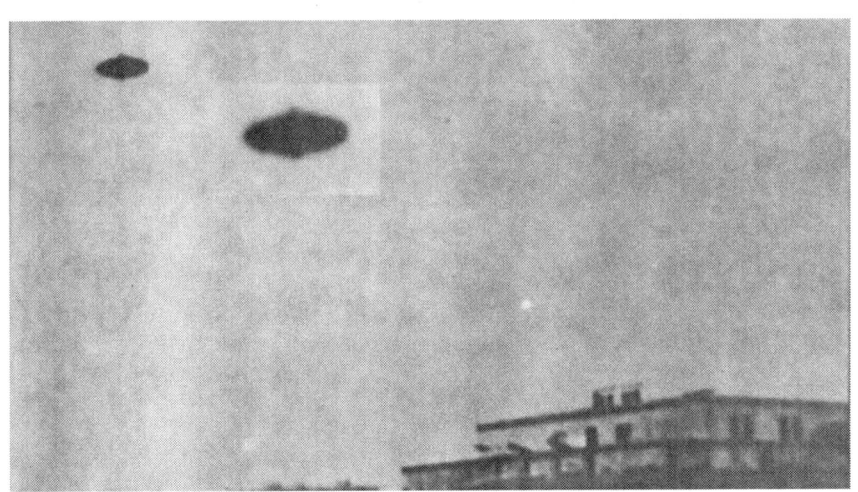

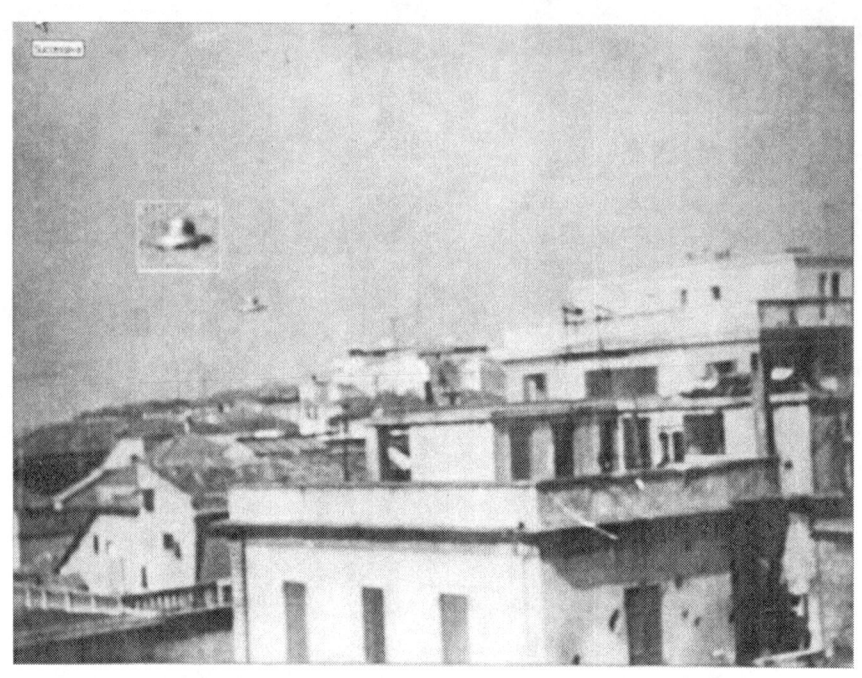

32÷35: Four pictures of a scout over Pescara, taken on September the 27th, 1957.

36÷37: Two spectacular formations of lights.

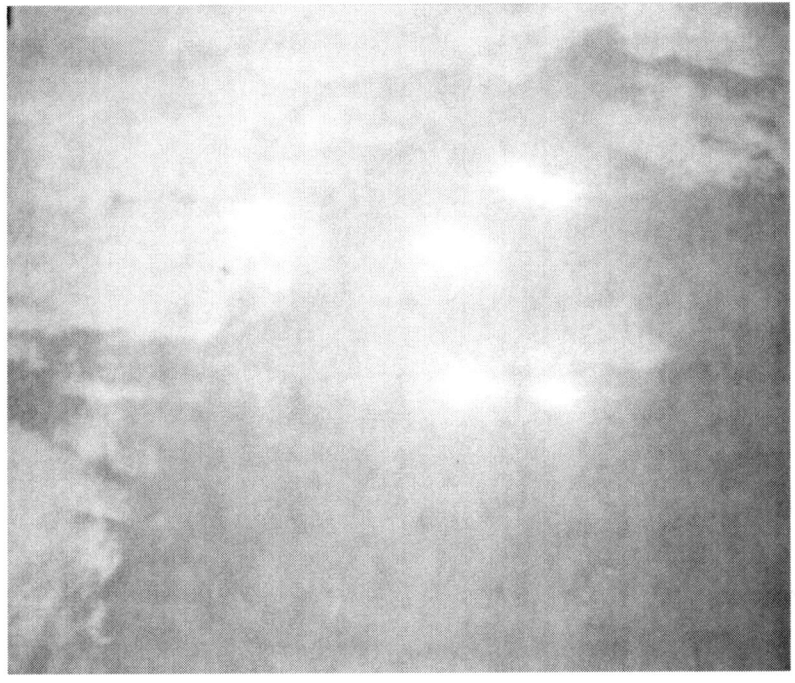

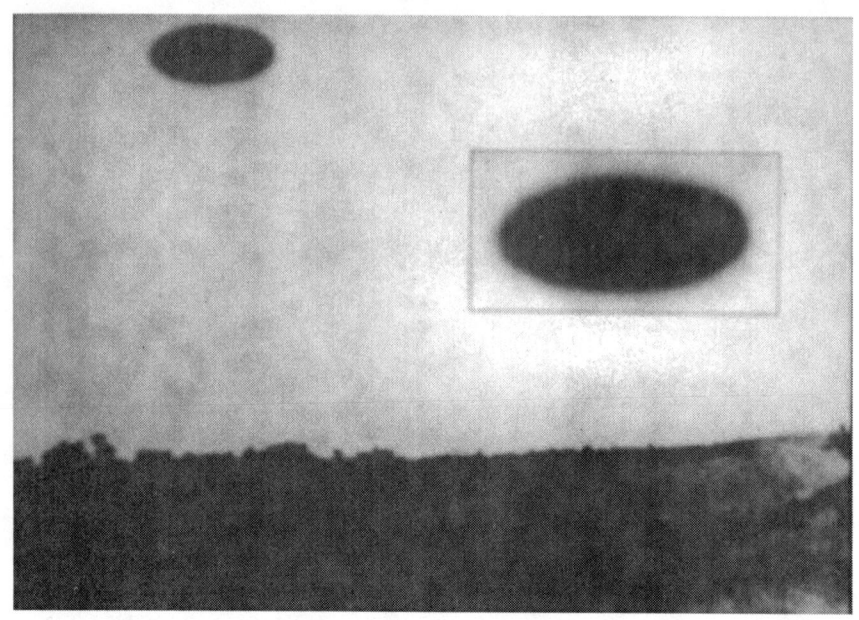

38÷39: Two among some pictures I've shot to a scout over some hills (summer 1964); now I live exactly in the point where these pictures had been taken from!

40-49: Two sequences of a low-flying scout; details about these pictures may be found in my comment within the report by Gaspare De Lama.

50: A scout over the shore of Montesilvano.

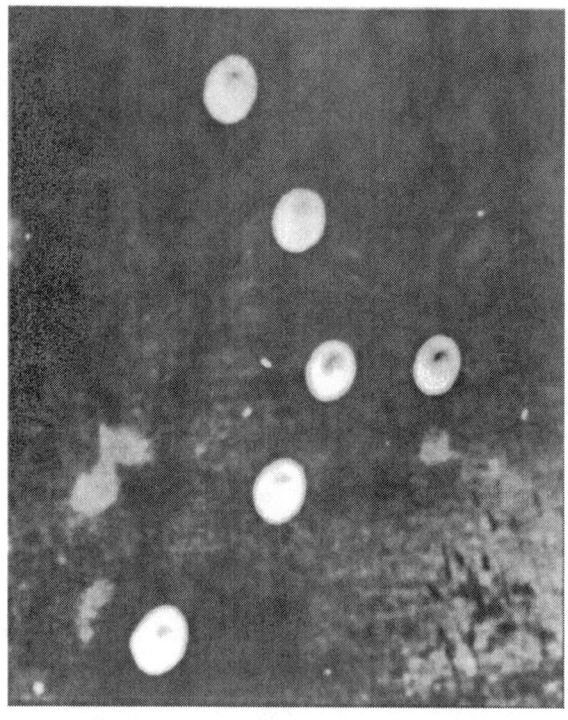

51: A formation of scouts, shot from a rather unusual point of view!

52: Pescara, October 1957.

53: Milano, Febrary the 12ve, 1962.

54: A mysterious picture, found by chance on Internet; probably it's a fake; it looks to show (one year in advance) the cross formation that Perego was going to see; the date is said to be December the 1st, 1953.

55: Passo Corese, summer 1971;
the <u>least</u> luminous object is the moon.

56: A scout, in an unknown place, at an unknown date.

57: A picture of Reiss Romoli, the company where I have been working in, then under construction; I have taken this picture around 1 p.m., with the shutter at 1/60; the place is Coppito, North of L'Aquila.

58: In front of Rocca Pia (Ascoli Piceno) at the times of these events.

59: In front of Rocca Pia, as it is to-day.

60: An unexpected image at the foot of a rampant of Rocca Pia; look at the plate of my car; light conditions (10 a.m., car in dim light) do not justify his phenomenon (October 2004).

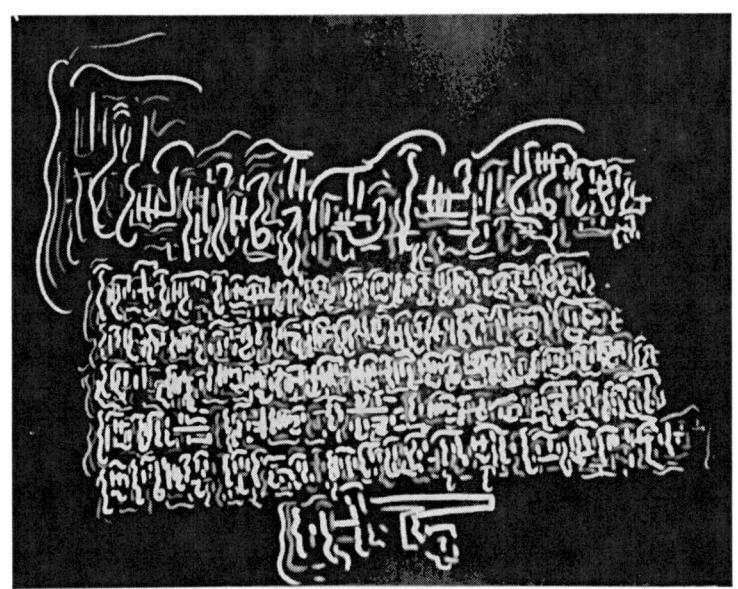

61: An example of their writing in the language common to all the people within the federation of the W56's. Colors have a semantic meaning.

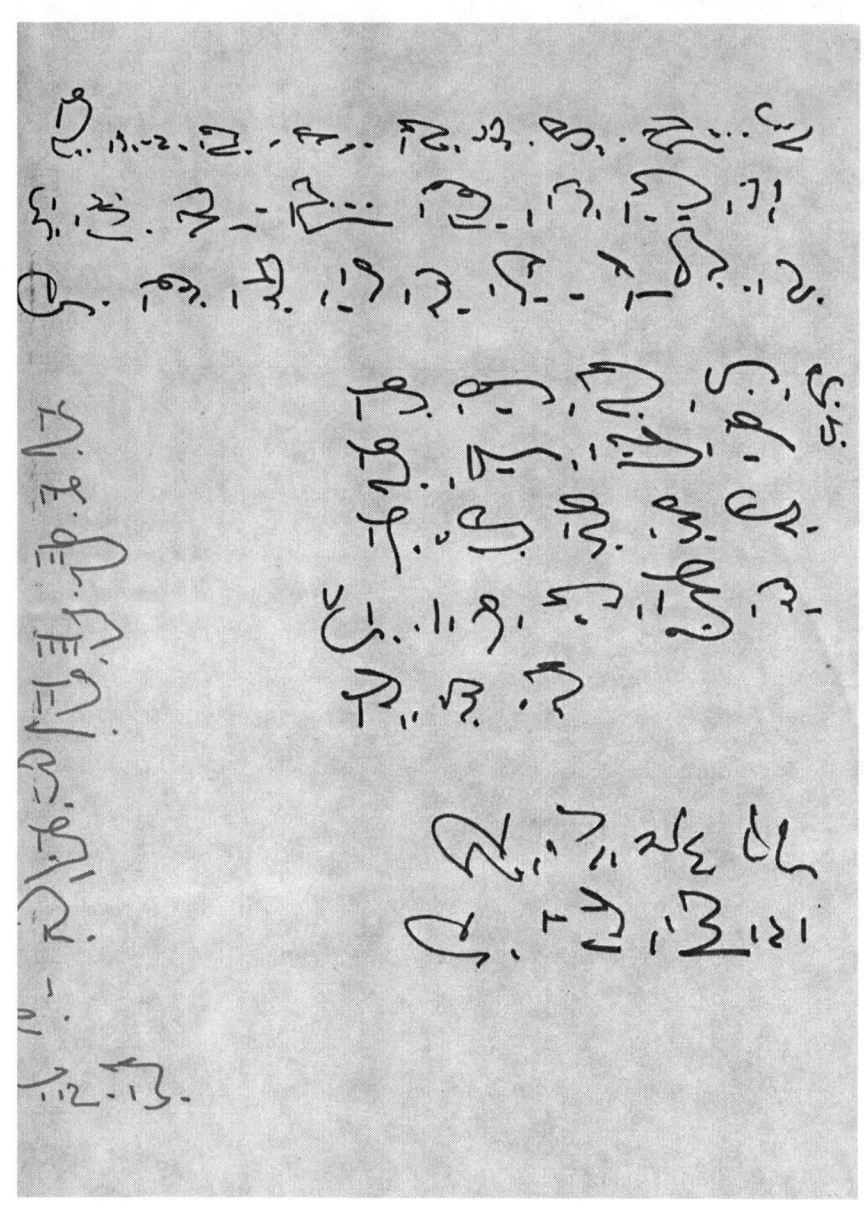

62: An example of one of the many "local" languages, spoken by a particular population; it looks that every symbol represents a word.

63: One of the devices built in those times, with the technology available. Expert of vacuum tubes will recognize two EL34 in the foreground, an 807 on the left (there were two of them), and a 6V6 in between.

64: As we have spoken so much about them,
one of the W56, in Bruno's garden.

65: A computer fake I made, a not too elaborated one; it required more than 600 FORTRAN lines; obviously all optical and lighting conditions have been respected. The object simulated in here is a strange scout, that has been seen many times in Abruzzo, once in France, and nowhere else, as far as I know.

66: The painter Gaspare De Lama, in a recent photo.

Curricula Vitae

In order that the reader may get a hint that we aren't totally mad, I'd like to present here two short *curricula vitae*, both of Bruno and of me.

Bruno Sammaciccia

Born in Ancona, April the 22nd, 1926. Degree in Psychology and in Psychiatry. He has been awarded four *Honoris Gratia* degrees, three International Academic awards, plus many cultural and theological awards (He had a trunk full of diplomas of every kind, and he couldn't make head nor tail of it). Has been a member of many academies, both Italian and foreign ones, and of several editorial boards. Has been very active in the spreading of moral, ethic and philosophic culture. A catholic writer, he has published hundreds of papers on related subjects, and since several years he has been writing also on foreign magazines. In order to increase his ethnological and philosophical culture, has visited many foreign countries: India, Central America, Middle East. His name has been quoted also by the Radio Vaticana (the Vatican broadcast), after some of the researches he has made in the fields of religions, magic, totem cults, myths.

In 1946 several magazines quoted the results of the researches he made following prof. Callegaris on the theory of cutaneous plaques.

Has cooperated to the cultural programs by UNESCO. In 1982 he has been awarded the title of *Man of the year* for culture.

He has written some 160 books, several translated into foreign languages, among which:

- "Religione e magia nei millenni" (Porziuncola, 1972);
- "The Secret Man" (Murrand, 1973);
- "O Milagre Eucarístico de Lanciano" (Mensagem de Fátima, 1973);

- "Il miracolo eucaristico di Lanciano" (Porziuncola, 1973);
- "Il vero Francesco d'Assisi" (Porziuncola, 1974);
- "La course vers l'infélicité" (Le Farté, 1975);
- "Alla ricerca di Dio" (Porziuncola, 1976);
- "Ho incontrato l'uomo" (Porziuncola,1976);
- "Eucharistic Miracle" (Kuba, 1976);
- "Le Miracle Eucharistique de Lanciano" (Éditions du Cèdre, 1977);
- "Trattato sulla Passiologia" (Fabiani, 1978);
- "Il volto santo di Gesù a Manoppello" (Porziuncola, 1978);
- "Das heilige Antlitz Jesu im Schweißtuch von Manoppello" (Porziuncola, 1978);
- "Le religioni primitive dell'Africa" (Fabiani, 1978);
- "False Myths of to-day" (Enderson, 1978);
- "Pedagogia scolastica" (Fabiani, 1978);
- "Meditazioni" (G. Fagiani, 1979);
- "Padre Domenico del Volto Santo" (G. Fagiani, 1979);
- "Beato Roberto da Salle" (Porziuncola, 1983);
- "Francesco, mio fratello" (Porziuncola, 1983);
- "La scalata della follia" (Porziuncola, 1983);
- "San Tommaso apostolo" (G. Fagiani, 1984).

He has sponsored the publication of more than 120 books about St. Francesco (for instance the first edition of his *Apocripha*) by the Porziuncola in Assisi; among them:

- Marino Bigaroni, ofm: "Compilatio Assisiensis" (Porziuncola, 1975);
- Giovanni Boccola ofm: "Opuscola sancti Francisci et scripta sanctae Clarae assisiensis" (Porziuncola, 1982);
- Marino Bigaroni ofm: "Speculum Perfectionis" (Porziuncola, 1983);
- Agostino Pensa, Luciano Canonici: "San Francesco, otto secoli di poesia" (Terni, 1985).

He has also been the subject of books written by journalists, or also just friends, for instance:

- Sadi Marhaba: "La via della realtà" (Porziuncola, Italy, 1975);
- Gaetano Tamborrino Orsini: "Bruno Sammaciccia: ventinove pensieri" (Porziuncola, Italy, 1977);
- Gianni Lussoso: "Bruno Sammaciccia, l'amico, il mistico, l'umanista, il pensatore" (Gira, Italy, 1980).

Hans

He had asked me not to divulge his data, and I comply.

Stefano Breccia

Born July 1st, 1945; degree in Electrotechnic Engineering at the University of Bologna, specialization in Telecommunications at the Politechnic University in Torino, specialization in Computer Sciences at the CNUCE in Pisa, specialization in Computer Graphics at the ICS in London, specialization in Company Management at the SSGRR in L'Aquila; has been working since 1973 at the Scuola Superiore Guglielmo Reiss Romoli (SSGRR), L'Aquila, where he has completed a full *cursus honorum*, working seven years as the Assistant to the Managing Director. Former external professor of Computer Sciences and Computer Graphics at the Engineering Faculty of the University of L'Aquila. Former external professor in charge of the course "Statistical Methodologies in Evaluating Training Results" at the Training Sciences Faculty of the University of L'Aquila. Former external professor in charge of the course "Algorithms for Computer Graphics" at the Engineering Faculty of L'Aquila. Has given lectures on didactical methodologies at the British Telecom training centre, on Fractal Analysis at the University of Novosibirsk and at the Soviet Academy of Sciences in Moskow, on the changes in roles because of Office Automation at Stratford upon Avon, within a Diebold project, on news in CAI methodologies in Madrid, within the framework of an IFTDO project, and so on.

Now retired, he has been in charge of the project for creating a post graduate TLC school in Cordoba, Argentina. Inside the framework of an European Community PHARE project, has worked in the design of TLC schools in Bucharest, Praha and Ljubljana. Within the framework of an EC TACIS project, has cooperated in the design and in the starting operations of a post graduate TLC school in Novosibirsk.

Has been President of the Scientific Committee of the "Centro Studi ed Applicazioni sulle Tecnologie dell'Informazione" (Catania). Former member of the Board of Directors of the "Centro Internazionale per le TLC Giovanni Someda" (Venezia), member of the Board of Trustees of the "Istituto per la formazione e la promozione economica" (L'Aquila), member of the Board of Directors of the SSGRR magazine "Società dell'informazione" (former Editing Director of the same magazine), member of the CT UNINFO Commission on Learning methodologies (Torino), and the Didactical Committee of the Italian Association for Industrial Research (Roma).

In 1985, has participated, as an EC consultant, to the Steering Committee of the DELTA Project, in Bruxelles, as the responsible of "Areas of Application of Artificial Intelligence Technologies in Computer Based Learning", and has participated to the DELTA Projects Selection Committee as the STET representative. Has participated, as an external consultant to the Italian Ministry of University and Scientific Research, to the definition of *curricula* in Engineering universities, particularly in the area of intermediate degrees. Hardware and software designer of the synesthetic project MC^4, exhibited at the 2000 edition of the SMAU, in Milan, in the area of Technology Innovation.

Has written the following books:

* Modulazione Delta Adattativa (1972, author);
* Metodologie di programmazione in FORTRAN (1982, co-author);
* Algoritmi per grafica basata su elaboratore (1988, author);
* I numeri nella storia dell'umanità (1995, author);
* Fractal: la teoria del caos (in progress - author);
* Algoritmi per Computer Graphics (2001 – author);
* Contattismi di massa (2007 – author).

and hundreds of papers and articles of specialized magazines.

End Notes

1. Father Enrico Sammarco, Passionist, in the foreword to the book "La via della realtà"; see Bibliograhy.
2. For those who might know Pescara: Bruno's house was in Genova street, and the timber warehouse was in a large (no longer existing) between Genova and Trieste street, near to the Milano street crossing.
3. How prophetic those words were to be!
4. The gentlemen were able to superimpose themselves to the usual broadcast, with an exceptional selectivity: putting together two different radio sets, tuned on the same frequency, only one of them was to receive the messages from our friends, while the second one was going on as usual.
5. One of the first small tape recorders on the Italian market.
6. Such objects were named "Aniae".
7. To them names mean nothing: all the names presented here have been attributed to them by Bruno or his friends, with very little exceptions (Dimpietro).
8. This funny event took place in a flat in Milano.
9. At those times, Forlimpopoli was a small village, some 30 km North of Rimini.
10. The aliens Bruno had named CTR's were the group hostile to the W56's.
11. Another of the W56's; the name had been chosen because, in those years, there was a popular radio broadcast, "Frosinone e Gallarate", a sort of parochial contest between the two villages.
12. Another friend, who was qualifying himself as an historian.
13. It should be evident, at this point, that our friends were living at ease within our environment: they were making use of cars, public bus service, and also of private planes.
14. The whole lenght of Pescara, some 10 kms.
15. I may confirm all that, after a lurid attack I've been subjected to, without even realizing the situation; when I "awoke", and reacted, everything stopped.
16. Extremely true.

17 In all evidence, a preposterous statement; may be they were willing to impress Bruno.
18 He is not meaning Paolo Di Girolamo.
19 Such performances always resulted in a violent blast.
20 Not really so, because, later on, pictures of those scouts started to circulate.
21 After so long a time has elapsed, this lady is still alive and well.
22 Sir E. A. Wallis Budge: "An Egyptian Hieroglyphic Dictionary" – Dover, New York, 1978.
23 Every amateur-astronomer may understand what Bruno is speaking about, substituting the word "galaxy" with "star cluster".
24 A physical entity that is, someway, immanent to both space and time.
25 This accident has been a very obscure one: the pilots were flying blind because of the weather, and they were on a totally wrong direction, 40° instead of the 332° required to land on Ciampino airport coming from South, but they looked to be confident in what they were doing, as if the two ADF receivers, and the three compasses, were giving erroneous readings, which has been confirmed by the subsequent enquiry.
26 Apuleius too has been a forerunner and, through a simple novelette has given us important suggestions: why the chief character, Lucius, has to eat roses at the end of the story? Moreover, in his "De magia" Apuleius gives us interesting hints on the mixtures between politicians and the religions, already at his times!
27 The distance along a quasi-constant latitude trajectory is 7,194 km, the one along a geodetic is 6,865 km, the distance through a straight tunnel is 6,538 km. To compute these numbers, I've referred to Fiumicino coordinates (41° 49', 12° 17') and JFK ones (40° 37', -73° 47').
28 Stefano speaking.
29 A third group, composed of people apparently above both the CTR's and the W56's, that, for operational purposes was

involving also earthlings, for instance an Austrian general who, in many cases, had taken care of Pescara events. I do not know what the probable acronym means.

30 The sun and the stars, in the XIV century are similar objects? Dante too must have been a W56 *ante litteram!*

31 Bruno speaking.

Bibliography

As I have already said, very little has been written about Amicizia. The titles I present here contain some veiled hints to this story. I do not quote the many newspaper articles, both because I cannot know all that have afforded this problem, and because, after so many years have passed, references have been lost.

Tullio Bosco: "Accadeva a Pescara" – Arte della Stampa, Pescara, 1992;
Paolo Di Girolamo: "Dossier UFO" – Mediterranee, Roma, 1980;
Paolo di Girolamo: "Noi e loro" (to be printed);
Ulrich Magin: "Von UFOs Entfürt" – Verlag C. H. Beck, München, 1991;
Sadi Marhaba: "La via della realtà" – Porziuncola, Assisi, 1975;
Alberto Perego: "Svelato il mistero dei dischi volanti" – privately published, Roma, 1957;
Alberto Perego: "Sono extraterrestri" – Alper, Roma, 1958;
Alberto Perego: "L'aviazione di altri pianeti opera tra noi" – CISAER Editions, Città di Castello, 1963;
Alberto Perego: "Gli extraterrestri sono tornati" – CISAER Editions, Lecco, 1970;
Solomon Shulman: "Инопланетяне Над Россией" – Эрмитаж, 1985.

As a gloss to this Bibliograhy, I'd like to quote ... a science fiction book:

Jimmy Guieu: "L'homme de l'espace" – Fleuve Noir, Paris, 1954;

Although it has been printed before 1956, and it's of course a work of fiction, it presents a context that is surprisingly similar to the one of Amicizia, with people for the Polar star fighting people coming from Deneb, to the point that, being one day in conversation with the Author, I asked him where he had got the inspiration from, but the guy didn't answer...